MONSTER OF THE MERE

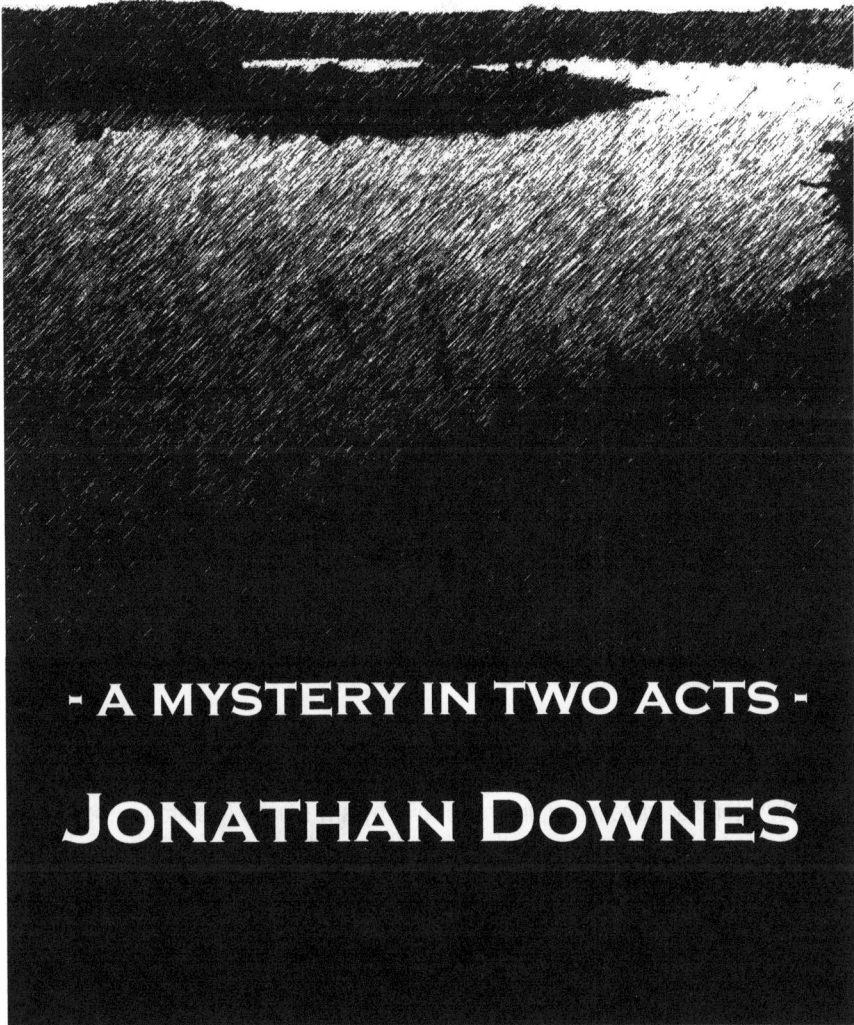

- A MYSTERY IN TWO ACTS -

JONATHAN DOWNES

Edited by Graham Inglis
Typeset by Mark North, Designed by
Mark North and Jon Downes for CFZ Communications
Using Microsoft Word 2000, Microsoft , Publisher 2000, Adobe Photoshop CS.

Photographs © 2002 CFZ except where noted
Maps © 2002 Graham Inglis/ CFZ except where noted

First published in Great Britain by CFZ Press

CFZ Press is a division of:

CFZ Communications
15 Holne Court,
Exwick,
Exeter.
EX4 2NA

© CFZ MMV -

ISBN: 0-9512872-2-2

- CONTENTS -

◾ ACKNOWLEDGEMENTS ◾

Many thanks to Tim and Lynda Matthews, Geoff Wright, Rob White-head, Karen Booker, Graham Inglis, John Fuller, Joyce Howarth, Steve Cummins, Mark North, Chris Moiser, my father, Bernie Starkie, Pat Wisniewski, the staff at Martin Mere nature reserve, Janet Walkey, Sarah Burdon, everyone at Champion Newspapers, Carl Lamb, The GMTV outside broadcast crew, Becky Weaver, all the eyewitnesses who were kind enough to allow us to record their testimony, my daughter Lisa, Andy and Cindy, David Curtis, all at LAPIS, Nichola Sullings....and of course Richard Freeman my friend and travelling companion for:

My mother Mary Downes (1922-2002)
And to the boys of the CFZ

It's dreamy weather we're on,
You wave your crooked wand,
Along an icy pond with a frozen moon.
A murder of silhouette crows I saw...
And the tears on my face,
And the skates on the pond they spell Alice

Tom Waits & Kathleen Brennan

- AUTHOR'S NOTE -

About half way through writing this book I realised that it had metamorphosed into something somewhat different than I had originally intended. This is often the way with my books: *The Rising of the Moon* started off as a socio-political examination of a 1997 UFO wave, and turned into an attempt to present a fortean field theory. *The Blackdown Mystery* started off as an examination of the socio-political impact of a UFO hoax, and ended up as a scurrilous novel. This book started off as a straightforward account of an investigation into a mysterious creature lurking in a lake in Lancashire. However it soon developed into the literary version of a fly on the wall TV documentary.

In October 2001, a student from Plymouth Art College approached the CFZ about doing a fly on the wall documentary about us. We agreed, and gave her complete access. She didn't bother to take us up on the offer, and although we haven't seen the finished product, I cannot imagine that the few sessions she filmed of me lecturing and chairing a meeting of the Exeter Strange Phenomena research group would be very edifying viewing. I hope that this book provides a warts-and-all look at the world's only full time cryptozoological research organisation as they celebrate their tenth anniversary....

‐ FORWARD ‐

I have met a lot of people in Forteana: some good, some bad, some sad and some distinctly and utterly mad. But I can honestly say that I have met none better than Jon Downes. Jon and I have been good mates since the latter part of 1997, when he interviewed me for the now late and lamented British newsstand UFO publication, *Sightings*. In the months that followed we would meet up at various gigs around the UK, such as at the Southend UFO Group's conference in 1998 (where we got wildly and spectacularly drunk and behaved *very* badly); the LAPIS bash in 1999 (where once again alcohol was an all-dominating factor and Jon's now-notorious "Stella experience" occurred); and at various and sundry lectures up and down the UK for a whole variety of groups dedicated to the study of all things unexplained. And from that heady and bygone year of 1997 when UFOs were forever-dominating (at least as far as the magazine shelves and the media were concerned), I would regularly travel to Jon's house for weeks at a time where, in the company of Jon's housemate, partner-in-crime, and lover of doom-laden music, Richard, we would horse around having absurd adventures in pursuit of "the truth" (which may or may not be "out there"), money and infamy. Not necessarily in that order, however. And we did a pretty good job on all three counts for three fun-packed years. But times change. In mid-2001 I traded in England for the far warmer and sunnier climes of Texas. But, via that marvel known as MSN, Jon and I keep in touch on a daily basis. And so when he asked me if I would like to write the *Introduction* to his new book, *The Monster of the Mere*, I did not hesitate to say "Yes!"

I had written the *Introduction* to one of Jon's previous books, *The Rising of the Moon* (which he co-authored with Nigel Wright), and a fine and

11

riotous read it was. And that's what I love about Jon's books. Unlike so many within Forteana, Jon has a fine and wicked sense of humour and a good understanding of the absurd. But he can also write and write damn well, too. Knowing how to weave a story together and how to create in the mind of the reader the feeling that they are right there, deep in the heart of the action, is one of Jon's great story-telling skills. In fact, I would say (without going way over the top with praise!) that Jon is probably the finest writer within Forteana today. And I do not make such claims lightly.

Take a look at his books *Only Fools and Goat Suckers*; the aforementioned *Rising of the Moon*; and *The Owlman and Others* and you'll hopefully see what I mean. If you don't, the loss is all yours.

Like all of those titles, *The Monster of the Mere* is a book that is at times delightful, funny, sad, insightful and adventurous. Coming across rather like the *Famous Five*-meets-*Fear and Loathing in Las Vegas*, the book should be read by anyone with even the remotest interest in cryptozoology or the study of mysterious and rare animals. It should also be read by anyone who wants to see how a Fortean investigation *should* be undertaken. Forget all those books put together by people who simply download all of their information from the Internet and produce the most sterile nonsense that one could imagine. Instead, focus your time and your hard-earned pennies on reading the words of a man who actually gets out there and does real research. *The Monster of the Mere* is not just a book that *can* be read; it's a book that *should* be read. And if, by the time that you finish turning the last page, your jaw isn't aching from laughter, then God help you.

Nick Redfern,
Nederfield, Texas
14 October 2002

OVERTURE
&
BEGINNERS

One wet and rainy afternoon in the middle of February 2002 a tired and bored reporter on a Lancashire newspaper finished typing up a news-story which he felt might be an interesting novelty item. As soon as he had finished, he ran it through his spellchecker, pressed "send" and the words on the screen in front of him were sent as a stream of electrons down the telephone line to the local news agency where someone subbed it, and then sent it round to various newspapers hoping that someone would pick up on the story and want to buy it.

Two days later, someone in cyberspace found the story. (S)he was a member of an electronic community known as *forteana@yahoogroups.com* and thought that the story might be of interest to other members of the group. Using the cursors on the computer keyboard (s)he copied the entire story, and by using the command Ctrl/v (s)he pasted it into an outgoing message template in her e-mail programme. Seconds later hundreds of people around the world had the story on the mainframe computer of their ISP waiting for them to download it.

The next morning a fat man with a beard woke up. He was in considerable pain, but then again, he always was these days. He had an impressive battery of physical and mental health problems, and to cap it all, the previous week he had been diagnosed with congestive heart failure and a few days later his mother had been admitted to hospital with what he knew, but could never admit to his father, was terminal cancer. He knew that she had only days or weeks to live and that soon the picture of his life would be changed forever.

The last thing that he was interested in that morning was unknown animals. However, this is what he did for a living; he was the Director of the Centre for Fortean Zoology - a non profit making organisation which was in the process of applying for charitable status. Despite his health and family problems there were still certain things he had to do.

He staggered downstairs to read the post. As he was leafing through the collection of bills, final demands, and letters from people all over the world seeking advice or information about mysterious creatures, he switched on his computer. As *Windows 98* fired up he sorted the post into three categories:

- Rubbish (This was immediately thrown away)

- Stuff that he could give to his assistant Graham to deal with (this went into Graham's filing tray)

- Stuff that despite the fact he was feeling like death, he had to deal with himself

As he started to write replies to the most urgent of the letters in group three, he logged on to the Internet and activated his *Outlook Express* e-mail programme. This morning there were 264 messages for him to deal with. He let them download as he finished writing his letters and then turned his attention to dealing with his electronic mail. There were a few personal letters from friends, most of whom had heard either about his dying mother or about his own poor health but most of it was rubbish. "Humorous" e-mails from total strangers which were about as funny as the Ebola virus. 'Spam' from various unknown sources mostly trying to sell him pornography, credit cards or cheap holidays, and a tranche of news items from various newsgroups that he was subscribed to.

He had got so used to sorting this stuff each day that he could now do it on autopilot. Most of the e-mails were rubbish, but as always, there were some music related news stories for his magazine *Baboon!*, some aquaculture related stories for his regular news column in *Tropical Fish* magazine, a few items about ornamental carp for his column in *Koi Carp* magazine and various cryptozoological news stories which he could use for the *Newsfile* section in *Animals & Men* the journal of the Centre for Fortean Zoology. Amongst the news stories that he filed away electronically for the latter category was one marked "Monster of the Mere". He

16

looked at it with cursory interest. Something was killing swans in a nature reserve in Lancashire. He shrugged his shoulders and filed it without bothering to read the rest.

All that interested him this morning was to take his morning medication, telephone the hospital in Barnstaple to see how his mother was, have a shower and go back to bed for a few hours. The Monster of the Mere was soon forgotten.

Seven weeks later, his mother had died and been cremated. At the same time as trying to find some time to grieve, the fat man with a beard was doing his best to put issue 26 of *Animals & Men* together in time for it to be on sale at the *Fortean Times Unconvention*. He discovered the 'Monster of the Mere' e-mail and included the following news story in the magazine:

On Valentine's Day the Liverpool Echo announced that the manager of the Martin Mere nature reserve in West Lancashire is convinced there is something large lurking in the deeps. Some creature, say staff at the reserve, is responsible for dragging fully grown swans into one of their lakes.

Several visitors witnessed the swan trying to flee the grasp of a giant underwater predator. In an earlier incident, the 20-acre lake where swans gather was left deserted as they all refused to go on to the water.

"Something is completely spooking them," commented reserve manager Chris Tomlinson.

"On two occasions, both Thursdays, January 17 and February 7, something in the water has caused the 1,500-plus wild wintering swans to completely disappear".

Centre manager Pat Wisniewski adds: "Whatever it was out there last night must have been pretty big to pull a swan back into the water. Swans weigh up to 13 kilos". Pat added: "This could be an extremely large pike, or a Wels catfish. Both conceivably could survive in the rather murky, de-oxygenated water for years and grow to an extremely large size". Four years ago, Pat spotted something that appeared to be the size of a small car circling the mere just below the waterline of the lake, which is four metres deep. One theory is that something may have

17

made its way into the mere through its drainage system many years ago as a juvenile and remained there ever since having grown too large to escape.

Over the next few weeks more information slowly trickled in. He spoke to Pat Wisniewski and decided that there was, indeed, something in the story worth following up. By the end of April the fat man and his colleague Richard, a balding plump Goth with a psychotic temper and a ridiculous sense of humour decided to try and investigate the mystery for themselves.

This is their story.

ACT ONE

Everything you can think of is true
The dish ran away with a spoon
Dig deep in your heart for the little red glow
We're decomposing as we go
Everything you can think of is true

Tom Waits & Kathleen Brennan

Wednesday 29th May 2002

I t's been a long time since I did this.

I have been a professional cryptozoologist and monster hunter for nearly half my adult life. It is a subject that has interested me for three and a half decades, and during those three and a half decades, I like to think that I have done a little bit to advance the sum total of human knowledge as regards the existence of animals that mainstream scientific belief systems choose, for whatever reason, to deny.

But it has never been quite like this before. The last time that I undertook an expedition was in January 1998, when together with my friend and colleague Graham Inglis, I travelled to Central America in search of the grotesque, vampiric, chupacabras. As I sit up on one elbow in my bedroom in a small seaside boarding house in Lancashire, I look at my reflection in the small, grubby mirror above the wash basin and realise, with sad resignation that the Jon Downes who drank his way bullishly around Puerto Rico and Mexico only four and a bit years ago, would find it hard to recognise the fat, disabled man with the shaggy hair and beard who sits typing on the bed in a dark room, illuminated only by the LED screen of his laptop as his friend and colleague, Richard Freeman snores contentedly a few feet away.

Four years ago I was far more idealistic than I am now. Scarified by the rigours of a horrific divorce, I took solace in the high adventure of a fear-

21

less vampire hunt (as chronicled in my book *Only Fools and Goat-suckers*) with as much gusto as I did my excesses of wine, women and song. For several years after I received the *decree nisi*, I generally behaved in an irresponsible and highly entertaining manner.

Four years later things are far more subdued. After the failure of another important emotional relationship during the dying days of the 20th Century, my mental and physical health (which had always been pretty ropey) began to deteriorate. In early 2001 my adopted daughter Lisa whom I love far more than I love anyone else on the planet returned home to live with me after a hiatus of some five years. With a daughter to look after and a growing realisation that it was probably impossible to achieve happiness through self indulgence in the sins of the flesh, I turned my attention to my life's work.

I had started the Centre for Fortean Zoology in 1992, more as a conceptual joke with a structure based upon what I had learned from Essex-based punk band and anarchist collective CRASS in the early 1980s. Now, with my newly rediscovered responsibilities, and my failing health which together with the death of my beloved mother in early 2002 made me more aware than ever of my own mortality, I realised that I would have to put the affairs of the Centre for Fortean Zoology on a proper footing.

As I lie semi prone in a seaside boarding house in Southport, drinking tap water out of an empty coke bottle and waiting for 20mg of valium to lull me into the arms of Morpheus, I can look back over the events of the last few months and honestly say that we are well on our way to achieving our aim. The once non-hierarchical anarchist organisation now has, at least a semblance of structure. We have employees, fundraisers and researchers, and before the end of the year, if Krishna wills it, we shall be a registered charity.

Four years ago, Graham and I drank our way through our Central American anabasis at Channel Four's expense. Today Richard and I are in a Lancashire Boarding House, preparing to hunt for the monster of Martin Mere, but we are sober, and everything to do with the expedition has either been paid for out of CFZ funds, or has been shoved on my Barclaycard. Like Clint Eastwood once said -

"We're back and this time it's serious" (or maybe it's just that at the age

of forty two I have finally grown up).

I have always had a predetermined concept of the North of England. During my days as a toiler in the rock and roll orchard I traversed the country dozens of times and spent much time in the decaying remains of what had once been the industrial heartland of this great country, until the twin evils of Thatcherism and cheap labour from the Pacific Rim finally put the kibosh on the dark satanic mills. These days it is no longer grim oop t'north, but even secure in this knowledge I was completely unprepared for the countryside which greeted me as I drove the elderly Jaguar that my father gave me only a week or so ago, from the M6 across country towards Southport.

The land is incredibly flat. For some reason I had always supposed the north of England to be hilly, but the land surrounding Ormskirk and Burscough is as flat and marshy as the Somerset Levels - an area particularly dear to my heart (as readers of my inky fingered scribblings will know). This is a strange terrain. Because I am writing a book about our adventures I would like to say that we were struck by a feeling of overwhelming foreboding as we crossed over the sections of the Leeds-Liverpool Canal which we know that we will have to explore over the next few days, but that would be a lie. The only thing strange about the countryside here is that it is completely different from what I had come to expect.

I said as much to Richard as we drove along merrily singing obscene doggerel to the tune of *Hi Ho Silver Lining* but despite Richard's totally tongue in cheek suggestion that I describe how we saw the long neck and hump of a typical Nessieform lake monster rearing out of the mist covered canal, there was no mist, no monster, and we only caught the briefest glimpse of the canal as we sped past. The only even slightly weird thing to happen during this leg of the journey is that the onboard trip computer (a name which has caused much merriment amongst those of my friends who remember my days as a long haired chemical dustbin) packed up for no apparent reason.

Even as I write this I realise that I am appearing horribly middle aged. This is probably because I *am* horribly middle aged. Although age has hopefully brought wisdom, I am only too aware that an account describing a middle aged disabled bloke with a credit card, expense account, mobile phone, portable computer and Jaguar driving in relative comfort around the North West of England may not have the immediate *Boy's Own Paper* appeal of my earlier adventures, in which drunk to hell I travelled around Puerto Rican hillsides lashed to the back of a Civil Defence truck as I sang Irish Rebel songs and tried to smuggle weird snails through customs. However I am older now, as (I suspect) are many of you reading this book, and anyway we've got a bloody big fish to catch.

As we drove into Southport itself in search of our digs, it soon became apparent that there was something else that we had not taken into consideration. Southport is a minor seaside resort. We had no insight whatsoever into the place or what makes it tick. As far as we were concerned it was just this place where our buddy Tim Matthews had his gaff, and from whence we had once obtained a tabby cat called Tommy.

We found our digs, and as arranged I telephoned Tim Matthews aka Agent Hepple, fortean, journalist, magazine editor, and alleged Government Agent, but there was no answer. As we knew that his lovely wife Lynda was due to have a baby any day now, I assumed that she may already have gone into labour and so we went off in search of something to eat. Half way through the meal Tim phoned and as it was nearly eleven and we were already well into our fourth pint we decided to write the rest of the evening off to experience and we made arrangements to meet on the morrow.

Many years ago my ex wife and I functioned as very minor cogs in the mechanism which promoted Chris de Burgh - a somewhat mawkish, though unimpeachably aristocratic - singer/songwriter with a ridiculous pair of eyebrows. As all the Chris de Burgh CDs went to Alison during the ritual carve up of our joint record collection during the divorce proceedings, and as I can't be bothered to replace them can't remember, I can't remember the exact song that I am about to quote, nor indeed the exact lyric, but a song from the album *Into the Light* extols the virtues of an "Out of season seaside town in the rain". Well, at the time of writing, it is just about out of season - for another day or two at least, and it was certainly raining, but the little corner of Southport that we have seen so far seems to be rather short on charm.

This is the first time that I have tried to write a book about an expedition 'on the hoof' as it were. After some soul searching I had purchased a cheap second hand laptop for 75 quid out of CFZ funds. This is so I could attempt to keep the narrative fresh and not be faced with the situation - like I was after our Central American junket - of trying to remember the details of what we had done, and how and why we had done it several months after the event. The downside is that I am writing stuff as it happens, and not with the benefit of hindsight, and I will be doing my best to avoid the temptation to adjust my ramblings retroactively.

The truth is that, although the parts of Southport we have seen so far are somewhat underwhelming the same could probably be said of the parts of any seaside town where in the second year of the third millennium one can still obtain bed and breakfast for fifteen quid a head. Like Chris de Burgh, although I hope that I have a little bit more style and élan than the bloke who screws his baby-sitter whilst his wife (the subject of his most popular - and most annoying song) is recovering from a broken neck, I too have a sneaking love for "out of season holiday towns in the rain". I did my nurse training in Dawlish, I have written a book featuring Exmouth and another featuring Eastbourne and Worthing. I have a sneaking suspicion that I will grow to - if not exactly love - have an uneasy and grudging respect for Southport in the same way that I have for all the other minor resorts that I have mentioned.

There is something quintessentially English about them in the same way as there is something quintessentially English about the Royal Family, and the pursuit of football. Throughout the day we have been confronted by cars, garages and pubs emblazoned with the cross of Saint George, either in support of Her Majesty's half century on the throne or, in most cases as part of some unspoken pagan ritual to try and ensure that David Buckram recovers from his foot injury in time for crucial games in the World Cup. There is something particularly surreal about chasing a monster in the heartland of England, during the World Cup, and over the Golden Jubilee weekend. It makes our quest seem a quintessentially English one. Maybe the "Monster" of Martin Mere is a 'Very English Monster'?

Despite my gastric rumblings - the aftermath of a large, particularly tasty and slightly indigestible Peking Chinese meal, the diazepam is beginning ever so slightly to take effect. Therefore, in the words of the greatest of all English diarists, (who was, unluckily for him, born several hundred years before the formation of the Roche Pharmaceutical Company of Geneva) *"and so to bed"*.

Thursday 30th May 2002

I always find it incredibly difficult to write a travelogue without sounding as if I am either aping Hunter S Thompson or Arthur Ransome. This is probably because the good doctor and the *"four children for whom the story was written in exchange for a pair of slippers"* have been my guides in travelling matters for all of my adult life. I have always vaguely wondered whether the Altoynan children (models for the Swallows) maintained their youthful *joi de* vivre into middle age, or whether, like me, they woke up grumpy and tousled, demanding tea and a shower before they would even speak to anyone. I know that Roger, the ship's boy, became a respected oncologist who ironically died of the very disease that he had dedicated his post *Swallows and* Amazons life to treating. I know that both Taqui (the model for captain John) and Titty spent much of the rest of their lives appearing at gatherings of fans who had dedicated their lives to the study of the books that chronicled their fictionalised childhoods. But I know very little about them as people. I suspect that they, like everyone else in the human race mellowed as they entered middle age - I strongly suspect that even Hunter Thompson began to want to start the day with a cup of Earl Grey rather than a hit of methadone as he got older. However, for many years *Swallows & Amazons* and *Fear and Loathing in Las Vegas* have been my twin inspirations, both in my writing style and in the way I have lived my life on the road. However, as I wrote last night, this time around it is different.

Quite how different it is this time I realised when I woke up this morning.

Richard woke up feeling ill with a sore throat. As I have already explained, my health is rapidly deteriorating, and I suddenly realised that if one or both of us were taken ill on this expedition there was nothing that we could do but soldier on. During my travels through the Mexican desert I had contracted a mildly unpleasant viral condition, and had been forced to lay low for three or four days, only emerging from a semi delirious state to perform like a trained monkey for the camera. However, I was surrounded by drivers and cameramen and assistants. There were eight or nine of us at every turn. Now there were only two, and I was the only one who can drive, so neither of us could allow ourselves to become ill.

Luckily it didn't seem serious and after a palatable, though slightly Spartan breakfast served by the peculiar couple who run the establishment at which we had stayed for the first night of our adventure, we seemed ready to start the day's events. The proprietors of the hotel were most peculiar. One struck Richard as being like a younger, sober version of my mentor Tony "Doc" Shiels. He had what would have been an elegantly sculptured beard and moustache if it hadn't been cut completely lopsided. He also had wild staring eyes and spoke in a detached, otherworldly manner. The other was a slightly aggressive Liverpudlian blonde who came across with the brashness of a scouse junkie as portrayed by H. P. Lovecraft. The guest house was undeniably strange, although they made us very welcome and comfortable.

Half way through breakfast my mobile telephone rang. It was Tim Matthews. The game was afoot.

As it happened, the day was a bright and clear one of the type which any of Ransome's young heroes and heroines would have immediately recognised. The only fly in the proverbial ointment was that no sooner had I typed the above paragraph announcing that the game was, indeed afoot, there was an ominous clinking sound and the mains adapter for my laptop gave up the ghost. This meant that we were not aware at this stage whether any form of electronic data storage would be feasible and we had our first interview nearly upon us.

Undeterred by these minor technical hitches we loaded our voluminous bundles of equipment into the Jag and set sail for the offices of the Champion Media Group. Almost immediately we discovered that we were driving in completely the wrong direction and as we made this slightly annoying decision it started to bucket down with rain of an intensity one only usually experiences in the tropics.

As we drove round and around Southport in search of the Champion Media Group offices I soon began to realise that I had sadly underestimated the town. It had some imposing and rather elegant architecture and although there seemed to be more than its fair share of distasteful looking thugs pacing the streets it seemed to be a town with quite a rich and varied cultural heritage. I made a promise to myself that as soon as I got back to Exeter I would do my best to find out some more about the history of the town.

It transpired that in back the Middle Ages, there were settlements at Ainsdale, Argameols, Ravenmeols and Morehouses (Hightown). However, Argameols had entirely disappeared by 1503, probably washed away by the sea during a period when the weather got gradually worse. The weather has always had catastrophic effects on the topography of the area – something that would prove to be very important as we explored the possible genesis of our quarry – and indeed the records are full of written accounts of terrible storms affecting the region in the early 18th Century.

It took the arrival of the railway in 1848 to make Southport a fashionable resort, although the pier wasn't constructed until 1860. However, somehow it managed to avoid the excesses of its neighbour across the Ribble - Blackpool - and to the present day is a far more pleasant place to visit. Southport, it seems always had slightly higher pretentious than did its sleazier neighbour.

By the middle of the 19th century, the Victorian passion for exotic flora and fauna had entered the home, via ferneries, aquaria, Wardian cases and conservatories. Part interior decoration, part education, part amusement, and very much a part of social life, these natural elements in the Victorian home lent a romantic air to the domestic scene. Even poorer homes could display pottery or picture frames decorated with flowers. The suburban greenhouse, a miniature version of those found in country houses and botanical gardens, acted as a conversation piece and social centre. Joseph Paxton's 1851 Crystal Palace offered a model for all, its initial use as an exhibition centre followed by its reconstruction in Sydenham in 1852-54 where it became a winter garden. It was the apotheosis of the iron and glass building, and its contents, the best of British industry, succeeded by the wonders of the natural world, enthralled much of the nation.

The success of the Crystal Palace encouraged Paxton in 1855 to propose a scheme for a glass-covered and tree-lined avenue circling London. It was to be called the Great Victorian Way, and was intended to provide an exotic environment for pedestrians and carriages. Owen Jones, designer of the colour scheme for the interior of the Crystal Palace, suggested a Palace of the People for Muswell Hill, London in 1858; it was to be a gigantic winter garden including a concert hall, museum, galleries and other attractions. Although these plans came to nothing, many smaller scale people's palaces, as winter gardens with an entertainment function came to be known, were proposed and sometimes built in the latter part of the 19th century. Iron and glass structures as adjuncts to houses and hotels, the conservatories and foyers, continued to be popular after the turn of the century as informal social settings, though the emphasis on plant life generally declined from its Victorian peak.

By the 1870s the winter garden was perceived as an ideal entertainment venue for a seaside resort, combining overtones of educational recreation with the possibility of using the space created for concerts and even dances. Large scale aquaria, too, were promoted at the seaside, often in conjunction with winter gardens. At the English and Welsh resorts, 40 limited liability companies were set up between 1864 and 1907 specifically to promote the construction or take-over of winter gardens, aquaria or both, and many other seaside entertainment companies also expressed interest in running similar attractions. The promotion of winter gardens and aquaria was most popular in the 1870s, when 22 of the 40 companies were incorporated, but only 11 winter gardens, some including aquaria, and five separate large aquaria were ever built, though many smaller ones undoubtedly existed.

Southport Winter Gardens was the first of the seaside winter gardens intended for the mass leisure market; opened on 16 September 1874, the sea front building was in the form of two pavilions connected by a covered promenade, and was designed by Maxwell & Tuke of Manchester. One of the two Germano-Gothic pavilions held a concert hall, the Pavilion, while the other was the iron and glass Winter Garden. The Pavilion had a capacity of 2,500 and the Winter Garden was 180 ft long with a maximum height of 80 ft. (In comparison, the central section of the Kew Palm House is just over 137 ft long and 63 ft high.) Southport Winter

Garden was advertised as the largest conservatory in England. Refreshment rooms and an aquarium filled the basement level.

Although it was one of the biggest edifices ever to grace an English sea front, the Winter Garden was not a great commercial success. The Southport Pavilion and Winter Gardens Company which promoted it was locally based and had wide support in the town and throughout the northwest with 350 shareholders at its peak, but the addition of Frank Matcham's Opera House on an adjacent site in 1890-91 strained the capital base too far, and the Company went into liquidation in 1898. Companies then rapidly came and went in an effort to make the Winter Gardens pay. Entertainments were driven downmarket; the Winter Garden was converted to a ballroom and roller skating rink, and the Pavilion became a cinema, but eventually both were demolished, the Winter Garden in 1933 and the Pavilion in 1962. Southport, although a rapidly growing resort, could not support an attraction needing such a large audience.

Southport Winter Gardens was also, at least according to the man we had come to meet, almost certainly the place mentioned by Frank Buckland in *Curiosities of Natural History*, which, thanks to the diligent detective work of our old pal Chris Moiser at Plymouth College of Further Education, provided the first link in the chain which may or may not link giant fish to the area surrounding Martin Mere.

Either by the standards of Victorian Eccentrics, or indeed by the standards of the motley crew who make up my own select band of heroes, Frank Buckland (1826-80) is/was/always will be a raging oddball. Son of the eminent geologist William Buckland, he set standards for the sensory appreciation of British wildlife, especially vertebrates, that have seldom been matched and never exceeded. Not content to confine his fascination to the more usual diagnostic characters (appearance, morphology, and the like), Buckland also set himself the challenge of demonstrating, not the mere edibility, but the actual palatability of British and exotic wildlife.

William Buckland set the stage for his son's lifelong appreciation of taste as a diagnostic character in natural history. Guests at the Buckland house might be, and were, served anything from mice to ostrich, along the way including hedgehog, puppy, crocodile and snails. According to Burgess, Richard Owen spent a queasy night after a dinner of roast ostrich at Buckland's, but John Ruskin regretted having missed the toasted mice.

31

William Buckland himself does not appear to have hesitated in the use of the sense of taste in analysing the natural world. Visiting a cathedral at which spots of saints' blood were said to be always fresh on the floor, never evaporating or vanishing, Dr. Buckland, with the use of his tongue, determined that the "blood" was in fact bats' urine. He mused in print as to whether the common mole or the blue-bottle fly tasted worse. His approach left quite an impression on young Frank.

Backhand's generation was concerned about the loss of agricultural land to commercial purposes in the UK and the resulting loss of domestically-produced animal food sources. The Society for the Acclimatisation of Animals in the UK, of which Buckland became a leader, investigated the feasibility of introducing various exotic animals into the UK for domestication as food sources. One Society dinner featured holothurian echinoderms ("strong in flavour, and excited a divided opinion"), kangaroo, lamb, wild boar, Syrian pig, and curassow.

Amongst the many species which the Acclimatisation Society attempted to introduce into the United Kingdom was the Wels or Sheetfish.

The wels, sheetfish or European catfish *(Silurus glanis)* is indigenous to continental Europe east of the river Rhine; it appears to be particularly common in eastern Europe, especially in the basin of the river Danube. It has a slimy, scaleless, elongated body and a broad, flat head with a wide mouth. Writing in *Naturalised Animals of the British Isles* Sir Christopher Lever notes that it has a "distinctly sinister appearance". He goes on to describe the creature in some depth:

European or Wels catfish *(Silurus glanis)*

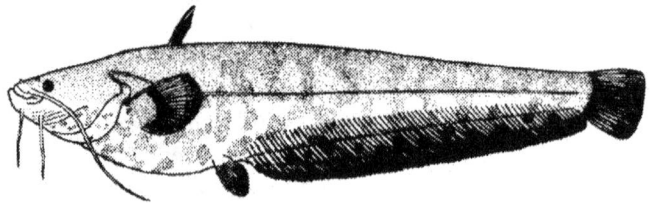

"The head, back and sides are usually some shade of greenish-black spotted with olive-green, and the underside is yellowy-white, with an indistinct blackish marbling; the head and back may sometimes be a deep velvety black and the sides occasionally take on a bronzy sheen. Two long barbels depending from the upper jaw, and four short ones from the lower jaw, help to give the catfish its name. There is no adipose fin, but an enormously elongated anal fin: a tiny dorsal fin is situated half-way between the bases of the pectoral and pelvic fins. The largest authenticated wels, taken from the river Dnieper in the Ukraine in the southern U.S.S.R., measured over 16 ft (5 m) in length and weighed 675 lb (306 kg); elsewhere in Europe and in England, however, the wels seldom exceeds 5 ft (152 cm) in length and 25 lb (I 1 kg) in weight."

The wels is a solitary fish, only meeting up in order to breed. It lives mostly in the still, deoxygenated waters of lakes, marshes and lagoons, but can also be found in the lower reaches and backwaters of slow-flowing rivers. Unusually for a catfish it is tolerant of both heavy industrial pollution and salt water and is found naturally in the brackish water of certain parts of the Black and Baltic Seas. Wels are nocturnal, choosing to feed after dark. They are voracious predators, especially when adults. Lever lists prey species as including: *"turbot, bream, crayfish, eels, frogs, roach, tench, ducklings, goslings and occasionally water-voles."*.

They have even been reported as man-eaters. Lever cites an authority called Valenciennes (presumably the eminent 19th Century Ichthyologist) as saying: *"In the year 1700, on the 3rd of July, a countryman took one near Thorn for Torun, [Poland] which had the entire body of an infant in its stomach'.* Lever also quotes someone called Grossinger (whom I have been unable to identify further) as saying that a Hungarian fisherman discovered the corpse of a woman in another *'having a marriage ring on her finger and a purse full of money at her girdle'.*

The name 'sheetfish', by the way, has been cited as being a corruption of "Shit-fish" implying that this voracious predator, which has also been described as a piscine vacuum cleaner, will eat absolutely everything including human waste. The wels usually spends the winter in a state of semi-hibernation, and emerges to spawn between May and July when the water has reached a minimum temperature of 68°F (20°C). Lever describes how

33

"the male scrapes a shallow hollow in the mud amidst thick vegetation close to the shore in which the female deposits about 15000 eggs per 1 lb (453 gm) of body-weight; these adhere to the vegetation which lines the nest, where they are guarded by the male until the young emerge after about three weeks."

He goes on to describe how in the eastern parts of its European range, especially in the vicinity of the Azov Sea on the borders of Russia and the Ukraine, the Caspian Sea, in Lake Aral (The Aral Sea) on the borders between Uzbekistan and Kazakhstan, and in those countries on either side of the river Danube, the wels is important economically. In eastern Europe it is stocked in a considerable number of commercial fish-farms (especially in Hungary) where its scaleless skin is employed in the production of glue and leather, and its eggs are sometimes used to "pad out" the more expensive eggs of the beluga sturgeon to produce an inferior type of caviar.

It was presumably this economic importance that inspired a succession of attempts to naturalise wels into British waters. The first known attempts to naturalise the wels in Britain took place in 1853 when, according to F. Day in his 1880 book *The Fishes of Great Britain and Ireland* quoted a certain Mr Llewellyn Lloyd who wrote: *'Through the indefatigable exertions of Mr George D. Berney of Morton, Norfolk, the silurus was last year [i.e. 1853] introduced into England. . .'*

We must agree with Lever who commented that "since nothing further is heard of this introduction, it must be assumed that it was a failure."

The second, and most successful introduction was described in *The Field* of the 7th September 1864 by James Lowe, joint honorary secretary with Frank Buckland of the Acclimatisation Society:

"That much desired fish, the Silurus, has at last been brought alive to this country, after various failures. The success is entirely due to the intelligent enterprise and perseverance of Sir Stephen B. Lakeman, who himself accompanied the fish all the way from Bucharest, a distance of 1,800 miles; and on Thursday night I had the pleasure of assisting Mr. Francis Francis (Piscatorial Director of the Acclimatisation Society) *in placing fourteen lively little baby-siluri in a pond not far from the fish-hatching apparatus belonging to the Acclimatisation Society on Mr. Francis's grounds at Twickenham.*

When I state that Sir S. Lakeman had to change railway carriages more than thirty times during the journey, not to mention other vehicles, such as horse-carriages and steamers; that he started on the 23rd of August, and arrived in London with the fish on the evening of the 15th of September; and that during all that long journey he had to wage perpetual battle with the indifference and stupidity of officials, from station-masters down to porters (most of whom seemed to regard the fact of his travelling with a strange fish as rather a misdemeanour than otherwise), the reader will have some notion of the difficulties which have been overcome.

The fourteen little siluri (or siluruses) which have arrived are what remain of thirty-six of the same species, which started from Kopacheni, where Sir S. Lakeman's estate is situated. This place is on the banks of the Argisch, a tributary of the Danube, and is about ten miles [16 km] from Bucharest. The Argisch abounds in silurus, and in all the other curious and almost unknown fish which swarm in the Danube, some of which (thanks to Sir S. Lakeman) we hope, at no very distant day, to reintroduce to their old friends, the siluri, in Mr. Francis's pond.

By way of preparation for the journey, the Siluri were placed in a water-cask, covered with a net, and placed in a large pond or lake of about 30 acres, belonging to Sir S. Lakeman; which pond abounds with fish, and yields silurus weighing up to 30 lb. and 40 lb. [13-18 kg], which may be caught with the line.... Of the thirty-six fish which started from Kopacheni, some were comparatively large (weighing up to 4 lb. [nearly 2 kg], and one of about 6 lb. [nearly 3 kg]), and some were mere fry.

Sir S. Lakeman started (as I have stated) from Kopacheni on Aug.23. He brought the fish, by Bucharest, to Giurgevo, a distance of fifty miles [80 km]; thence by steamer to Basias (in Transylvania), and so on by railway to Pesth, Vienna, Nuremburg, Cologne, Brussels, and Boulogne. The larger fish died first, all but the six-pounder, which endured to Vienna; and he only died there, it is supposed, because the servant in charge put his barrel into a stable, and it is likely that the ammoniacal atmosphere of the place disagreed with him.

On arriving at Folkestone, there were fourteen survivors of the thirty-six which started from Kopacheni, and I am happy to say that every one of these reached Mr. Francis in the most lively and promising state. .

Immediately on his arrival in London, Sir Stephen Lakeman, with most praiseworthy public spirit, thought more of the fish than of himself; for without even driving to an hotel, he made his way to The Field office; and I need not describe with what delight he and his charge were welcomed. In a very short time we were on our way to Twickenham.

So we got them safely down to Mr. Francis's, and on the brink of the pond turned them into a trough - fourteen little siluri, all alive and kicking, and as spry and frisky as possible. Their size varied from an ounce and a half to two ounces [42-56 g], for they are not more than three months old; but Sir S. Lakeman (who is well acquainted with the fish) declares that in a few weeks, when they have had the benefit of fresh water and plenty of food, their increase will be rapid and astonishing. When put into the water, they dived down to the bottom at once, with an easy vigorous movement, and waving their long barbels about, quite as if they knew their way about the pond which they then saw for the first time. From their flourishing condition, there is every reason to hope that they will increase and multiply. Indeed, I have now very little doubt that (with ordinary luck) this country has now acquired the Silurus Glanis. This is (so far as I am aware) the first time that this valuable fish has been brought to our shores; and the gratitude not only of the Acclimatisation Society, but of the country, is due to Sir Stephen B. Lakeman for the admirable manner in which he has effected the task which he unselfishly (and let me say patriotically) imposed upon himself

Soon afterwards, Buckland *'took down ten of them to my friend, Higford Burr, Esq.,.; Aldermaston Park, Reading, and turned them out into a large pond in front of the house. Some three years afterwards this pond was let dry - the silurus had entirely disappeared.'*

This is the largest and the most successful introduction of wels to Britain, although all authorities agree that there were several other importations throughout the 19th Century including fish from The Danube introduced to Woburn Abbey in Buckinghamshire. Today they are found in several isolated colonies in the south of England, especially in the Buckinghamshire area and in the River Ouse.

What, however, led us to believe that Frank Buckland or his cronies had introduced these remarkable fish to Lancashire?

I make no apologies for quoting the following passage in all its glory. It is not only important, because it places Frank Buckland squarely in Southport at the time of the heyday of the Winter Gardens Aquarium, it shows that he visited the said aquarium, was on remarkably friendly terms with its proprietor (although as H.M. Inspector of Fisheries, and a well known popular science writer he would probably have been welcome anywhere), but it confirms that Southport Winter Gardens and Aquarium was a remarkably sophisticated establishment, which would almost certainly have had one or more wels catfish amongst its exhibits. However, my real reason for quoting the tale of "Uncle Tom the Alligator at the Southport Aquarium" (from his posthumous book Notes and Jottings from Animal Life which was published in 1882) is that it provides a charming insight into the mind of a man who is not only one of my great heroes, but who is – if Richard's and my theory is correct – one of the most important and pivotal figures in this current investigation:

"I was much pleased to be able to pay a short visit to the Southport Aquarium on my way home from the north in November 1876. The most remarkable novelty was a splendid alligator about eight feet long. He had a nice spacious glass cage all to himself, the two ends of the cage containing gravel stones, and in the centre was a pond in which he could bathe when so inclined. When first bought, Mr. Alligator was in a very seedy condition indeed, terribly thin and wan looking - in fact, half starved. His skin was all in cracks, and his coat of mail had to be oiled every morning by means of a flannel on the end of a stick. This acted like a Turkish bath to our friend, and did his constitution good. For many days, even weeks, he sulkily refused to eat, and lay quiet and still like a stuffed thing. At last he took all of a gulp a live pigeon, and ever afterwards he fed well. The secret of getting him to eat was temperature - temperature the old story, the key to so many fishery problems, whether of salmon, oysters, or alligators. Hot-water pipes were introduced under the floor of his den, and Mr. Alligator, feeling the agreeable heat to his gouty toes and elegant I figure, fancied, I suppose, he was back again in the tropics, so he woke up and began to eat; and what more tasteful beginning could there be than a nice live pigeon with feather sauce?

This ogre now feeds capitally on pigeons; in fact, he is getting expensive to keep; be will eat beef, fish, and almost anything. They don't stand live pigeon dinners every day.

Anxious to show me his pet feeding, the curator offered to give the alligator more supper; he had already devoured his proper supper. The cu-

rator got on to the top of the cage and touched him gently with an iron rod. I was surprised to see the activity of the rascal; he opened his eyes with a jerk, up went his head like a run-away hansom cab horse, he gave an indignant whisk with his tail like a lady picking up her skirts when a clumsy fellow puts his foot on the pet lace, and, to my surprise, began. to puff himself up. Gradually he became larger, larger, and larger, like the blowing up of a football; his armour glittered, and the bony studs stood well out from the soft intermediate skin. I confess, when at his full I longed to run a pin into him to save his life, as I saw he had a chance of meeting with the same fate as the foolish frog in Aesop's Fables, who vainly puffed himself up trying to become as big as the ox, with whom he was having an argument. Just, however, as he came to the bursting point, Rex alligatorum suddenly relaxed himself, and his steam escaped, I suppose, through his larynx and nose. Anyhow, he began a most sonorous hiss. 'H-i-s-s, h-i-s-s;'

I can hear it now - just the noise a dragon ought to make. It was like no hiss I ever heard before, much deeper and louder sounding than any snake. As he continued his hissing he became thinner and thinner, till he looked quite the skeleton of his former pretty self. Then he began to blow himself up again, for (I could see it) the iron rod was getting up the monkey of Mr. Alligator.

A chicken's head and neck were then suddenly thrown into the bath; in an instant Leviathan forgot his rage. (Mem.: when a Homo bipes implumes, one of our own noble species, loses his temper, give him a dinner, and he will be all right, showing once again that 'the nearest way to the heart is down the mouth.') However, hearing the chicken's head fall splash into the water, the alligator - he should be called Uncle Tom - was after it in an instant, and seized it just as a dog catches up a running rat in his mouth.

He first of all bit it spitefully as though to kill it, if it happened to be a live thing; and then - one, two, three, and away - chump; back went his head, down his throat went the chicken's head in a moment. Reader, hold your nose and swallow a pill before the looking-glass, and you will understand how Uncle Tom swallowed the chicken's head. His blackship then gave a gulp, and, like the 'Oh the poor workhouse boy' in the song, asked for more. Three chickens' heads and a bit of beef, extra rations, did Uncle Tom get that evening, and all on my account. Supper over, he crawled on to his warm bed of shingle, and as the door over his head closed, he lazily shut his eyes, as much as to say, 'Thank ye, my boy, you may come as often as you like. now don't bother me, I'm going to sleep; good night, my hearties.' 'Here, hie! please give me a nice live pike for my breakfast to-morrow morning. I like pike; I shall dream of pike, for I like just as you like a bit of sport as well as a bit of grub, and if I can combine the two, why, so much the better.

Really, 'Uncle Tom' is a grand beast; he is growing so fast that he is to have a new drawing-room and dining-room, and then he will have space to swish his tail; he has not much room for his tail just now. I wonder how it is that in the 'struggle for existence' his tail has not begun to curl; may be his descendants in one hundred thousand years will have their tails curled up like a pugdog's. By the way, why do some pigs wear straight tails, some curly tails? There's a problem for you.

Besides big Uncle Tom, there are a number of smaller alligators. Close to the end of the Uncle's cage is a charming family of baby alligators, from ten inches to one foot long. These little boys and girls have a nice hot nursery, heated from underneath, and a flannel blanket over their dear little heads. They are as active as blackbeetles, and when their counterpane is taken off, scuttle away in all directions. If I reckoned right, there were twenty or thirty of these little fellows. Several of the ladies in Southport have purchased pets from among them, and it may be that no Southport lady will consider her establishment perfect without a baby alligator to bask on the hearth-rug, and go out for a walk on the promenade with her. When the pet defunets, he can be stuffed, gilt, and put in the hat for an ornament, don't you know?

However, those in the Aquarium are growing fast; they gorge like charity children at a 'tea and bun' festival. The keeper cuts up fish into small bits, and throws them into the cage; they scramble for them famously, and apparently love each other in that disinterested, charitable, and unselfish manner which may be seen by a careful observer who throws down handfuls of coppers among the London gamin and street Arabs in the crowd when waiting for the Lord Mayor to pass through Fleet Street on 'All Sprats' Day, November 9.

But there are yet more of Uncle Tom's relations at Southport. A huge box, looking like a gigantic coal-scuttle, stands near the boilers in the engine-room. Open sesame! and lo and behold a nest of young alligators of all sizes and shapes, like the ladies' bonnets and hats in a Regent

39

Street shop!

The curator dives his hand in and picks them out one by one, holding them aloft like an old fishwife in the Edinburgh market selling Scotch haddies. The lot are not yet presentable. They have not yet received the certificates of the School Board, and their tempers and appetites are not sufficiently mollified by the furnace fire to go into the glass apartments which are getting ready for them, so they remain at their ease, toast themselves before the engine-room fire, while the engine-driver consoles their minds by whistling to them' Tommy, make room for your uncle,' and feeds them with bread and cheese, which they will not eat."

The implication is that the aquarium at Southport actually managed to breed the American Alligator – a feat which is notoriously difficult even today and which suggests that, despite the dubious ethics of flogging off the surplus stock as pets to the ladies of Southport, the institution in question had very high standards of excellence in animal husbandry.

It would be surprising if an institution of this degree of sophistication had not kept, and quite possibly bred, wels catfish.

However, back to May 2002.
We did, eventually manage to find our destination, and as I deftly squeezed the Jag into the only available parking space (which turned out later to belong to the Managing Director), I could foresee serried ranks of newspaper employees lined up in horror at the thought that not only had their managing director's parking space been summarily taken, it had been taken by a scruffy looking bloke who looked a little like George Harrison would have done in 1972 had Krishna insisted upon a diet of doughnuts rather than brown rice for his followers, and a plump, balding Goth dressed in black and sporting a leather waistcoat and a black silk shirt.

Such people as us, we began to realise, don't really exist in Southport.

Deciding to blithely ignore this fact we entered the building with as much panache and élan as we could muster. This is surprisingly difficult when you can hardly walk and have to prop yourself up with a stick. However, I did my best, and I think that I acquitted myself reasonably well. Richard, as always, swaggered about the reception area as if he owned the place, and when we spied an old fashioned brass bell which was emblazoned with the words "RING FOR ATTENTION" we did what we were told and rang it.

A spotty youth with a vacant expression answered our summons and looked blankly at us when we explained that we had an appointment with Geoff Wright, the Chief Reporter of the *Skelmersdale and Ormskirk Champion.* A few minutes later Geoff arrived. a cheerful, balding man of uncertain age, he impressed us immediately with his knowledge of local history and his energy and very real interest in or quest. He apologised for only being able to give us a few minutes of his time that day, but because of the imminent Jubilee Bank Holiday, and the ensuing chaos in newspapers across the land, his deadlines were completely awry, but, he said, we could all meet up for lunch the next day.

He did, however, give us a photocopy of the first news item that his paper had run on the subject of what we had dubbed "The Monster of the Mere". It was dated 13th February 2002 and read:

WHAT LIES BENEATH?

Terrified Swans forced to flee the grasp of giant underwater predator - it could resemble something like this, say wildlife bosses at Burscough Reserve....

This accompanied a very blurry picture of what looks to me like a South American Red Tailed Catfish - a species which is unfortunately kept fairly commonly by aquarists despite its reputation as a "Tankbuster". These magnificent fish can grow up to five or six feet in length and can be apparently as intelligent as a particularly stupid dog. However they are creatures of tropical waters and would die within days if released into the murky waters of Martin Mere.

The story continued...

"You've heard of terrors from the deep such as Jaws and the Loch Ness Monster, but keep your eyes peeled for a new beast on the block - the Beast of Martin Mere.

Until now, wardens at the Burscough reserve have been keeping tales of a colossal beast lurking in the depths very much to themselves. Such sightings include reports of a dark shadow "the size of a small car" seen

41

moving beneath the water's surface. But these stories were given a new twist this week when wardens became convinced the birds were being "spooked" by something sinister....

These opening paragraphs are a fortean's dream. The story gives you JUST enough information to spark off an investigation such as the one which we are now embarked on. SOMETHING is lurking in the DEPTHS of a mysterious body of water - the mystery is tremendous and the technique of invoking both the Loch Ness Monster and the eponymous elasmobranch of *Jaws* is a masterpiece of journalism.

One's appetite well and truly whetted for a magnificent monster hunt, one reads on with mounting excitement...

Whole flocks have been taking to the sky, often disappearing only to return later and resettle on the lake. Now, a horrific incident witnessed by several people last week has confirmed that something very strange is indeed going on. Visitors looking out over the floodlit waters on Thursday night were shocked by the sight of a swan trying to flee the grasp of a giant underwater predator.

According to witness reports, all was quiet when a panicked swan was seen struggling to take off....

The skilful journalism is building to an orgasmic climax. Mike Warren, the geezer who wrote the story, has introduced a new twist to the tale. Like the spectators in the Roman arena, we, the readers are enthralled witnesses to a gripping saga of animal torture and death. No matter how hard we pretend to be civilised inhabitants of the third millennium, the spectacle of an innocent creature in its death throes is irresistible to many, and better than sex to some.

The story continues:

The brave bird managed to fight its way to the lake shore and was just scrambling out when the voracious beast dragged it back in. Although the swan escaped and recovered after a good rest, those who witnessed the attack were shaken by what they saw.

Even more than a gruesome tale of animal brutality, the English love nothing more than a heart-warming tale of animal (or human) bravery, and the English sympathy for the underdog (or in this case the underswan) is encapsulated perfectly in this story. Then, in the final part of the account, the author sets out the *dénouement* - the fact that those in command are taking the situation seriously and are determined to do something about the matter:

Now wardens have vowed to find out just what is lurking in the murky depths of the mere, and if possible take it alive. Speaking after the drama, Centre Manager Pat Wisniewski said: "Whatever it was out there last night must have been pretty big to pull a swan back into the water. One of our wardens had witnessed something a bit similar a couple of weeks ago when all the swans suddenly cleared off the main mere. I must admit that the birds have been a lot more nervous this year.

Then came the double whammy - the first hand testimonial...

"About four years ago I noticed something very large on the margins of the mere. Obviously because the water is so murky I couldn't get an idea of what it was but it seemed to produce quite a wake.

Wardens are currently baffled as to what is living in the water and could have grown to such a size.

But Pat admitted he "likes a good puzzle" and is intent on finding the answer. He added "I'm a biologist so I tend to look for rational answers. It could be a very large pike but another creature which has regularly turned up in lakes in Britain is the Wels catfish. Both, conceivably could survive in the rather murky de-oxygenated water for years and grow to an extremely large size.

Wels catfish can grow extremely big with heads and mouths the size of dustbins; they have been reputed to swallow dogs and will eat more or less anything. They live in very sluggish muddy waters, often in deep holes in the bottom of a lake or river and basically they are omnivorous, they will grab anything they can get their mouths around...

After having described his favourite suspect, (which is, by the way the one that the CFZ team also favour) Pat reiterated his desire to solve the puzzle:

I would quite like to get to the bottom of it myself as I quite like a good puzzle. I have never seen the birds reacting in the way that they are at the moment. Perhaps when the wild birds aren't there in such numbers we might make an attempt to net the lake and see what we can haul out.

We use a huge net a bit like a trawler, it's almost that sort of system. We put the net on a boat and have a several man team haul it across to the shore - that's the only chance I see of getting it....."

The piece concludes with some facts and figures about Wels catfish.

The species is of perennial interest throughout Europe. I recently submitted the following news item as part of my job as News Editor for *Tropical Fish* Magazine:

DING DONG BELL – DOGGY'S IN THE WELS

The Wells, or European catfish is one of the largest freshwater fishes in the world, and in recent months has been causing havoc across Europe. On the 18th October 2001 news agencies reported that there was an ongoing hunt in Germany for a giant catfish which had eaten a pensioner's paddling dog. Dozens of volunteers were reported to be willing to fish round the clock to catch it before it went into winter hibernation. Apparently the catfish grabbed a dachshund puppy belonging to an elderly woman as it paddles and swam a few metres from the lake shore. However, like so many of these stories, nothing ever gets heard from it again, and so news section of Tropical Fish magazine is unable to print the conclusion of this strange tale There is no doubt, however, that the wels is native to Germany. However, it is a little known fact that it was introduced to parts of the UK in the 19th Century by the massively eccentric Frank Buckland founder of the Acclimatisation Society whose main raison d'être was to introduce exotic creatures into the British countryside and then eat them. It seems that even now there are a few left because in early July 2001 it was announced that "The hunt is on for a giant catfish which is posing a risk to the native fish population of a Kent river".

The fish, which for some obscure reason was named 'Darren', a 40lb Wels catfish, was caught and then released by 15-year-old angler Oliver Parker-Grater on the River Darent. Jo Hunt, of the southern regions office of the Environment Agency, admitted that not only is the fish a major threat to the native brown trout population in the Darent but that it had no natural predator to control its population growth.

44

Ms Hunt said the fish is unlikely to be killed. "It is part of the environment and nature and that is what we are trying to preserve," she said.

She predicted Darren will meet a similar end to another catfish, Hannibal the Cannibal, who was housed at Brighton Sealife Centre after being caught in a Sussex lake earlier in the year, but on the 26th July, Adrian Saunders, who had been leading the search for Darren, today said: "We have spent the last three days using electro-shock fishing, but we have been unable to locate him. My suspicion is that he has swum downriver towards the Dartford Creek area and could even be in the River Thames."

And once again nothing more was heard. It seems that these huge fish are just as socially elusive as their more famous giant relatives in places like Loch Ness. There are always plenty of news reports, but very little hard evidence!

After having read the cutting and admired the draft of another article, this time written by Geoff himself, in which he describes the advent of the CFZ team to the fray, and which was illustrated by a flattering and rather funny cartoon, we bade our leave of Geoff, first asking him to tell our old friend Tim Matthews that we were in reception ready to meet him for lunch.

Tim came bounding downstairs to meet us.

He is one of my closest friends in the bizarre business of fortean journalism, and he is also one its most enigmatic and controversial figures. His shadowy past has been linked with ultra right wing political groups, covert military intelligence, political chicanery and skulduggery, football hooliganism and a long standing feud with self-styled bastion of left wing morality Larry O'Hara. He has been accused of everything from membership of the ultra right wing terrorist group Combat 18 to being an establishment spy sent to "destabilise" British UFOlogy.

As, by and large, British UFOlogy is the last refuge of many people who are so eccentric that their behaviour borders on, and in some cases goes far beyond total insanity, the idea that the British Government (or indeed any other Government) has the slightest interest in destabilising it is frankly laughable. Whatever Tim is, or appears to be, he is one of my dearest friends, and we cheerfully shook hands, and went out to the car park where I gingerly drove the Jag out of the parking space belonging to the Managing Director, and made our way to a spectacularly insalubrious pub where I bought lunch for the three of us.

We talked of this and that, and although our conversation was cheerful and wide ranging, neither politics or our current quest were mentioned. Although we speak most weeks on the telephone our paths do not meet up that regularly and so when we do occasionally meet in the flesh we talk about things of far more importance to both of us than who rules the country or what is swimming in Martin Mere.

The pub was festooned with Union Jacks, celebrating both the Golden Jubilee and the World Cup. It was only later that I realised I had missed a wonderful chance of taking the best Tim Matthews portrait ever conceived. The man himself surrounded by symbols of British nationalism. However I didn't think of it at the time and we ate a particularly inedible meal and chatted cheerfully.

After lunch we bade Tim farewell. He clicked his heels and left us as we made our way through the twisting Southport Streets towards the open countryside where we planned to have our first sight of Martin Mere itself.

As we left the fleshpots of Southport behind us we saw a Comet Superstore by the side of the road, and pulled in to try and see if we could find out what was wrong with my laptop. The very fact that you are reading this book now is testament to the fact that it turned out that all that had gone wrong was a blown fuse in the plug. Ironically we were served by a cheerful scouse lad called Dave who had seen Richard and me making our unsteady way back to our digs from the Chinese Restaurant the previous evening. He admired the Jag, and asked what we were doing in the area. Upon hearing that I was writing a book set in the area he demanded to be given a glowing mention, preferably involving a close encounter with a glamorous and pneumatic young lady.

Well Dave, you've got half your wish. Although I can't really introduce a non-existent blonde bird with big tits, I can tell the public at large (or at least the portion of them who buy, and hopefully read my books) that this book at least would not have been written without your help!

We then drove out towards Burscough.

As we drove, Richard read me Geoff's article about us:

Attempt to solve the "Fear in the Mere" mystery

THE MONSTER CATCHERS ARE COMING

EXCLUSIVE

by Geoff Wright

Professional 'monster catchers' are heading for West Lancashire - in search of what is laying beneath the troubled waters of Martin Mere. In February THE CHAMPION reported on terrified swans being forced to flee the grasp of what appeared to be a giant underwater predator - possibly a massive Wels catfish - the biggest species of freshwater fish in the world!

Two internationally acclaimed big fish hunters (professional zoologists) are heading this way to study the situation, and then return in July in the hope of nabbing a record-breaking terror from the deep. Such a find would make national if not international news putting the spotlight on Martin Mere...

Richard gasped. Neither of us could even slightly be described as "big fish hunters" and we had no intention whatsoever of "nabbing" the creature. We merely wanted to photograph it, establish its identity and then return it to the wild. As for "putting the spotlight on Martin Mere" we didn't know for sure until we spoke to Pat Wisniewski on Monday, but we suspected that this was the very last thing that either he or we would have wanted.

Richard continued to read me the article:

The Wels catfish is without doubt one of the Ull's most fascinating fish.; it is not a newly introduced species but has been in English waters for over a century - but very few have actually been caught. The dynamic duo - Jon Downes and Richard Freeman - will bring specialist equipment and a desire to seize the colossal creature that's causing mayhem in the usually tranquil waters of the Burscough reserve - therefore the new beast on the block could have its days numbered.

Richard looked at me in dismay. We had no specialist equipment apart from a couple of digital cameras. The whole point of this leg of the investigation was to make friends with the folk at the Mere and not to jump in with both feet making extravagant claims that we were not in any position to substantiate. We certainly had no intention of killing the creature, which is what the story implied. The rest of the article reiterated Pat's comments in the February piece that is quoted above.

We decided that the most important thing that we could do would be to go to the Mere as quickly as possible and try a preliminary recce before our Monday rendezvous with Pat. We had originally planned to spend some of the afternoon in the local history section of Burscough Library but with the possibility of a storm breaking over our heads in the aftermath of Geoff's well meaning attempts to give us some extra publicity we felt that we should change our plans.

Driving through the small, and surprisingly unprepossessing town of Burscough with some difficulty we found the road to Martin Mere. Once again we marvelled at how flat the countryside was. Although we were undoubtedly in West Lancashire the land was as flat as coastal Lincolnshire, and was criss crossed with an intricate network of canals, drainage ditches, streams and other waterways, which gave the landscape a bizarre patchwork effect.

As we finally drove into the open gates of Martin Mere nature reserve we were somewhat taken aback by how big it was. We had seen maps on their web site, but somehow we had envisaged a small network of interconnected ponds with a few ducks, not what seemed to be a major tourist attraction costing over a fiver a head to get in. However, we paid our fivers and went in. Purchasing a map of the complex we ambled gently past a pond featuring a bizarre mix of Gosanders and flamingos, towards the path which led to the Mere itself.

This was our second surprise. For some reason I, at least, had imagined a lake with relatively untrammelled access to all. I guess that Richard had envisaged much the same thing because he had made a point of changing his slightly dandified footwear for a pair of utilitarian Doc Martens before we left the car park. However, to our surprise, the lake, or at least that part of it accessible from the path that led from the Visitor's Centre was surrounded by an earthwork some fifteen or twenty feet high.

48

It was not until we had followed the path around to the first of the bird watching hides that surround the lake that we had our third surprise. Grateful for a place to rest as my back and legs were hurting like billy-oh I collapsed into one of the sturdy wooden benches provided for the comfort of the visiting birdwatchers. Richard did the same and we both took our first look at the Mere.

It was about a tenth of the size that we had been led to expect.

In the original news story which we had plundered or the NEWSFILE of *Animals & Men* #26 we had been told that the Mere itself was some twenty acres in area. Just a cursory look proved that it was no such thing. A rough estimate based on the size of my father's garden back in North Devon suggested that it was, perhaps, two and a half acres in area at the most, and peppered with tiny islands, each of which seemed to be populated by nesting waterfowl. So much for our preconceptions of a huge expanse of water upon which the birds were frightened to swim.

Coots, moorhens and whole families of ducklings did their own inimitable thing, and the lake itself looked remarkably placid. We took a few photographs and then wandered along to the next hide where we got into a conversation with a guy called Bernie Starkey who ran a small shop selling telescopes and binoculars to the visiting bird enthusiasts.

Gingerly we broached the subject of the Monster of the Mere, and to our surprise he was happy to talk about it - mainly because he was extremely sceptical on the whole issue.

Whilst not for a moment doubting the word of either Pat or the other warden who claimed to have seen a huge creature in the lake, he quite reasonably questioned the relevance of the "attack" on the swan. After all, he pointed out, in the winter the waters are seething with available prey species. Why should any predator risk attacking anything as potentially dangerous as a fully grown swan.

Richard and I have both had run ins with these birds over the years, and were forced, despite our disappointment, to agree. Then Bernie dropped his bombshell. He had, so he told us, a perfectly workable theory to explain the "attack". The previous winter at about the time that the incident was reported, at least one of the swans which had overwintered on the Mere had been suffering from an unspecified neurological disorder

49

which made it behave very much like a cow with BSE. The symptoms of this "mad duck disease" were simple. Although the bird seemed perfectly normal whilst it was on the land, as soon as it entered the water it was unable either to swim or indeed to maintain its equilibrium.

As Bernie pointed out, the sight of a huge and panicky swan rolled over on its side and kicking its legs and wings about in a vain attempt to try and get back to the safety of dry land, could well appear - to the casual observer - as if something was attempting to drag the poor unfortunate fowl beneath the surface of the lake.

As Bernie was the first to admit, this scenario didn't explain the sightings of an object "the size of a car" nor the mysterious "bow wave" seen by Pat. As Bernie was at pains to stress, both witnesses were extremely experienced naturalists who were almost certain not to mistake a known animal for an unexplained phenomenon. Also, Bernie admitted, none of this explained the mysterious behaviour of the waterfowl over the previous winter. They had indeed been unsettled and had behaved in a bizarre manner. There might well be a mysterious beast in the lake, concluded Bernie, but he didn't think that it had been attacking swans. After having heard his testimony we were bound to agree with him.

We bade Bernie farewell, grabbed a quick cup of tea in the Visitors Centre cafe and then made our way back to the car for our journey to Blackpool - to the abode of my old mucker Rob Whitehead and his long suffering girlfriend Karen (who had never actually met us before and wasn't really prepared for the Centre for Fortean Zoology Expeditionary Force to take over her house).

When we arrived they both made us feel comfortable and welcome and I settled down to attempt to write some deathless prose before going out to the pub for the weekly meeting of LAPIS (Lancashire Aerial Phenomena Investigation Society) where Richard was to sing for our collective supper.

We went to the pub where we spent as convivial evening meeting old friends and making new ones, and where I spent far too much money on alcohol and Richard delivered an entertaining lecture on the links between Dragons and UFOs before we staggered back to Rob's place where he and Richard (and another LAPISite called John) drank, watched television and guffawed loudly at each others jokes as I retired to the dining room to finish writing up the day's adventures. By the time I had finished Richard and Rob had retired to bed, and the house was silent. As I type these final words about the day's events I can only guess at what trials and tribulations tomorrow will bring. However, if the investigation continues in the same vein as it has so far it will have been worth it.

Friday 31st May 2002

A word is in order now, I think, about my relationship with the Captain and Crew of the Good Ship LAPIS. The reasons for my brief involvement with UFOlogy are well documented. Back in the balmy days of 1996 there was a brief spurt of media interest in UFOlogy, which coincided not only with the 50th anniversary of the Roswell incident and the birth of modern UFOlogy, but it also coincided with my divorce. As a result of the media furore surrounding the whole subject there were at least half a dozen magazines devoted to things saucerlike and flying.

Well, I have never made any secret of the fact that I am no more than tangentially interested in the subject of folk from Outer Space. I am a fortean zoologist pure and simple, and although there are certain sectors of the two disciplines which overlap, (most notably cattle mutilations and the predations of the chupacabra), most of the subject of UFOlogy leaves me cold, and I have never shown more than the most passing interest in it. However, I had an expensive divorce to pay for, and needs must when Adamski drives, so I went into W.H. Smiths and bought copies of each of the then currently available magazines on the subject. Over the next few days I telephoned each of the editors, presenting myself as an acknowledged expert on the subject (when, of course, I was no such thing), and managed to secure enough commissioned articles to ensure that I was able to purchase my wife's share in the equity of my house.

Over the years I have written a number of books which have been marketed to a UFOlogical audience. Only one, a tiny tome called *UFOs over Devon*, was actually a UFO book. Despite the flying saucer on the cover, *The Rising of the Moon* was an investigation into fortean field theory, and *The Blackdown Mystery* was a semi-fictionalised account of an investigation my team of investigators carried out into an incident which, it turned out, never actually happened. It is the nearest thing to a novel I have written to date, and it was mostly an excuse to lampoon my closest and dearest friends in print.

As a result of these books I found myself ensconced on the UFO Group Lecture Circuit, and although I have never really made any attempt to hide my lack of interest in the more hard-core elements of UFOlogy, I have found myself as a regular guest at meetings and conferences around the UK. In the late summer of 1999 I received a telephone call from a geezer called Robert Whitehead who asked me whether I would like to appear at his annual conference at Lytham St Anne's in Lancashire. As there was a hundred quid in it for me, as well as my expenses and two nights in a hotel I agreed readily, and so in December, my co-author of *The Rising of the Moon* a skinny Jewish geezer called Nigel Wright and I set sail up the M6.

A week before our journey north Rob telephoned me again and asked whether I had any technical requirements for my talk. I have always been a firm believer that if a lecturer can't get his point across without the help of visual aids and technical jiggery pokery he is a poor sap indeed and so I declined his kind offer and, jokingly said, that the only thing that I would require was a four pack of Stella Artois lager.

This was a mistake, because Rob, obviously doing his best to cut this southern interloper down to size interrupted my talk on the Saturday morning by coming up onto the stage with a four pack of Stella for me. There was no way that as a scholar and a gentleman that I could let this pass, and so, in the full view of the assembled throng I drank the lot as I concluded my talk. I then proceeded to spend the rest of the weekend in a drunken haze which was lampooned by Andy Roberts in his scurrilous news-sheet *"The Armchair UFOlogist"*: I am very fond of Andy, who is undeniably one of the wittiest men in UFOlogy. *The Armchair UFOlogist* is an occasional publication in which, since 1997ish Andy has lampooned the great and good (and the small and bad) of the subject in an insightful and amusing manner. He had made quite a few enemies (and a smaller number of close friends) through the publication of what some people still insist in describing as a 'squalid little rag'. Many people are furious when they are lampooned therein - others, like me and Nick Redfern are disappointed when we're not.

At the LAPIS conference in 1999 Andy began by describing the events of the Saturday daytime:

"First sight I saw was Jon Downes trying to flog his pathetic wares from behind a desk which was clearly out of proportion to his size. He regaled me with tales of the goings on at the hotel on the previous night...... clearly I had missed something as observing ufologists in their social interactions is far more interesting than any stupid stories they hawk round the conference circuit.

Jon gives a new meaning to the phrase 'largeing it' and is clearly ufology's answer to Blackadder's Bishop of Bath & Wells. But who has the drawings? All say in a breathy voice "Dear boy, I don't give a flying fuck what you think of me" and rightly so Jon. Great guy, and the only other person in UFOlogy besides Neil Nixon who you can have a cracking conversation about music with. He knows his stuff. Surely by now there should be some form of test whereby if you don't know who, say Mighty Baby are, you aren't allowed to be a UFOlogist?

Straight into Jon and Nigel Wright's lecture. This was based on their book Rising of the Moon, a Keelesque romp through various 'paranormal' and UFOlogical events in the south west a couple of years ago. Fascinating I'm sure, but it seemed to me like they'd just stitched any old cobblers together and made an adventure story out of it. Whats more the only visual aid present was the arrival on stage of a four pack of Stella.

Apparently Jon had been asked if he needed any visual aids and he (in jest) requested said four pack, The audience gasped at the sheer audacity of the man. Jon just drunk his breakfast and rambled on. Somewhere on the front row you could see a hunched figure in a parka cringing. A thought bubble appeared above his head saying "preposterous, that would never happen at a YUFOS conference and the fat bastard would have to wear a suit anyway", yes it was Graham Birdsall, well out of his natural habitat at someone else's UFO event. Bet he didn't pay for the ticket!

55

[NB: The excerpts from *The Armchair UFOlogist #4* are used by the kind permission of Senor Roberts, but because I am an egotistical so and so, I have edited his narrative severely so as only to contain those bits that are about me and which explain, like nothing else can, my deep, meaningful and complicated relationship with the LAPIS posse]

Andy continued...

"But what's this? Lurking at the end of the foyer, looking ever so slightly nervous, was Tim Matthews. Tim had to make an appearance because well, because, that's what he does, just so people don't forget him. We'll have much Tim talk later but for now all you need to know is he was handing out leaflets for some rally or other and acting furtive with Downes - who would have to be in very deep cover indeed for anyone not to spot him
.

Luckily I caught them flagrantly giving Nazi salutes and I sincerely hope someone uses this as proof that they were there and I had a camera."

Admittedly that was all vaguely true, but I was rather drunk and there was a perfectly reasonable explanation for our actions. I just can't remember what it was.

Andy then accompanied the rest of the party, the half mile or so from that year's venue - a seafront pavilion and concert hall in Lytham St Anne's - to our hotel; The Edenfield. This is a most peculiar venue - it is apparently owned by some friendly society of retired professional upholsterers in the North West. It is a strange mixture of Victorian elegance, art deco grooviness and horribly tacky exhibitions of horribly tacky upholstery. There is what seems to be a permanent display of mass produced teddy bears in the foyer; each of them grinning at each arriving guest with a spastic artificial fibre smile. It is a weird place, but their salmon mornay is excellent and they serve the most fantastic claret that I have ever drunk.

According to their website:

"Our traditional Lounge Bar is perfect for pre dinner drinks, and once you've polished off your delicious four-course meal, why not return to sit back and relax in comfortable surroundings to reminisce over the day's activities and make plans for tomorrow"

56

Once a year or so, the elegant surroundings are completely overwhelmed by events such as that evening's as described by Andy R:

".........we all decanted into the piano room for a few hours of the most bizarre UFOlogical post gig 'fun' I have been present at, and I've been at a few. Simple Beatles songs soon gave way to rock standards belted out by a man who I spoke to much but know only as Dave from Geordie-land, aided and abetted by a freeflowing permutation of Jon Downes, Nigel Wright, Sir Malcolm of Robinson, and many others including Miss Bott on backing vocals. Louie Louie, Stand By Me, all the UFOlogical classics were trotted out and then it was into Irish rebel songs such as the touching version of The Armagh Sniper delivered by Jon (bar bill for the night £65.00) Downes, now doing a passable imitation of Citizen Caned......the most responsible of us such as Posh UFOlogist, Nick Red-fern, merely looked on in disbelief........Matthew Williams skulked in an earnest fashion and then went off to ring his mummy.......we were joined again by the Hull people one of whom was well oiled and confided to all and sundry that he was a bouncer, and kept showing parts of his anat-omy whilst questioning the availability of leeches for it. Clearly I was missing the UFOlogical context he was getting at here and his friends eventually took him away.

It's perhaps best to ignore Jon Downes' frequent and desperate pleas to Irene Bott for something called 'executive relief'. Thankfully Irene is far too expensive for Jon to merit even a look of disdain. Just because he's a media whore and arts editor of the Planet on Sunday (didn't Clark Kent work for them?) it doesn't mean to say he can get away with this sort of behaviour.

So, that was LAPIS 1999, it was good and you should have been there. LAPIS also hold the distinction of holding the last UFO conference this millennium (depending, of course on when you believe the next millen-nium starts)."

Since then Rob and I have been close friends. Richard and I appeared at the 2001 conference where even more alcohol fuelled high jinks ensued. The LAPIS conference, is, we believe, the best one of its kind in the UK, and is the one which provides the bench mark by which we measure our own annual event - The Weird Weekend.

Until this particular expedition our only encounters with Senor Whitehead and his band of Merry Men have been fuelled by massive amounts of alcohol and insanity. I have a vague memory of a drunken conga line rambling around the Edenfield Hotel at Lytham St Anne's at four in the morning, during which the great and good of British UFOlogy sang a chorus of Tom Waits's song "Singapore".

We had assumed from our boozy encounters with the man that Rob, like us, lived in some degree of bohemian squalor. We had no compunction, therefore, in inviting ourselves to stay for four days whilst we carried out our investigations into the Monster of the Mere. When we got to the house where he lives with his charming girlfriend we were shocked to find that we had invited ourselves to encamp in a desirable and very elegant residence and all my visions of sex drugs and rock and roll flew out the window. I said this to Karen as we had a drink together on the Friday night and she found the concept uproariously funny!

Today, however, we had a considerable amount of research to do, and only a very limited time in which to do it.

Today was also the day when the whole face of the investigation began to change.

Today was also the day that everything began to go horribly wrong.

The day started well enough. It was another feast of sunshine and blue skies, and much to our joint pride Richard and I managed to find our way out of Blackpool without mishap. On the outskirts of Preston we filled up with petrol and were pleasantly surprised to find that despite its reputation as a gas-guzzling giant, the Jag was actually doing something in the region of thirty miles to the gallon which was actually quite heartening . However the journey from Preston to Southport took considerably longer today, and we were feeling rather flustered as we drove towards the town for our rendezvous with Geoff Wright.

We telephoned him to see what time and where he wanted to meet, and the first blow of the day happened. He had, so he told us apologetically, forgotten his glasses, and so he would have to spend his lunchtime rushing home to get them rather than seeing us and helping us with our investigations. "Never mind," we said, secretly thinking that this was perhaps the most bollocks explanation that we had ever come across in a long professional history of being given bollocks explanations. However, he had been helpful on the previous day, and it would be churlish of us to complain, we thought, and so we decided to check out the little town of Burscough.

Burscough is the nearest conurbation to the Mere itself, and we had been told (I forget by who) that it had an excellent local history section. Without much difficulty we found the library, and discovered to our pleasure that yes, it did indeed, have a disabled car park. Things were looking good, very good, in fact until we actually went inside and discovered that not only was there no reference section at all, merely two tiny tables with uncomfortable chairs stuck bang in the middle of one of the aisles, but there was no photocopier and there was some noisy and particularly annoying refurbishment work going on which seemed to necessitate one having to research against the background of the sort of noise (or power tools and assorted carpenters) that one would normally expect from a building site.

Feeling somewhat annoyed, we decided to swallow our ire, and get on with the job in hand and therefore we sat down to commence what we could of our research. One of the most important things that we wanted to find out was the origin of the name Ormskirk. Orm (or Orme) is the old Norse for Dragon and almost without exception every place name which includes the name ORM, ORME or WORM (for the root of the word ORM is the same as that of those segmented little fellows who live in the earth of your garden) is associated with a Dragon legend, and furthermore usually an aquatic dragon legend.

For the next little section I am indebted to Richard. There is an old adage that one does not keep a dog and bark ones-self. I would update this in cryptoinvestigative terminology and say that it is absurd to share a house with the world's most knowledgeable dragon expert and even attempt to explain about dragons for yourself.

RICHARD FREEMAN: The orm or lind-orm is the most prevalent of Scandinavian dragons. It resembles a gigantic snake and is often said to bear a crest or mane like appendage along its neck. Lindorms were believed to start life as ordinary snakes. Over centuries they grew to such vast sizes that they took up residence in lake were the water helped support their massive coils. As time drew on they out grew even the lakes and had to move out to sea. Thus during their enormous life spans Lindorms became both lake monsters and sea serpents.

59

The most infamous lindorm was the Jourmungandr or Midgaurd serpent. This monster was so long it encircled the whole world in its coils. The Jourmungandr kills Thor the thunder god in Ragnarock, the demise of the Norse pantheon.

The breath of the lindorm was not fire but a cloud of noxious gas that was said to be so potent that it could shrivel crops and strike down birds in flight. The fear of these creatures lasted into Christian times when Lindorms were meant to encircle churches and crush them. Villagers bred gigantic bulls to do battle with the lake dwelling serpents, a combat that usually ended in the demise of both participants. This piece of folk-lore has an interesting modern analogue. In South America giant anacon-das are sometimes referred to as "manatoro" [the killer of bulls] as it is believed huge specimens can constrict and swallow fully grown bulls.

In Britain the lindorm can be found in such legends as the Lambton Worm, the Nunnington Worm and many others. These poison breathed, limbless giants are distinct from the more familiar winged, fire breathing dragons.

Even today the lindorm persists many deep Scandinavian lake are said to be homes to such creatures. These include Lake Storjsen and Lake Sel-jord. Lindorms have even been filmed and photographed, if a bit blurrily. Eyewitnesses now number in the thousands and the creatures have even been seen crawling across the land.

It seemed more than a little coincidental to us that the very site of one of the more intriguing zoological mysteries of our times should be in the precise location of a place whose very name suggested that it had draco-nian connections, and although we were still convinced that the animal for which we were searching was flesh and blood in nature, and probably a giant wels catfish, as fortean zoologists in the truest sense of the word, we were determined to leave no stone unturned in our search for the truth. Therefore, our most important quest at Burscough public library was to discover whether or not Ormskirk, had, or ever had its own Orm.

There was indeed a tiny local history section in Burscough Library. It consisted of about four short shelves of books, and included some ex-traordinary inclusions such as the biographies of Keanu Reeves, June Whitfield and Larry Grayson, and an in depth history of the venerable TV Soap Opera *Eastenders*. However in between these paens to the more banal end of popular culture were some excellent books on local history, amongst which was a self published tome dated 1968 which told the story of the history of Ormskirk.

Reasonably excitedly I whispered to Richard that I had found something. However, because of the disgusting level of background noise, the whisper was inaudible, and I tried again at slightly louder pitches until we were actually conversing in voices which by normal standards would have been considered uncomfortably loud. However, because at this point the sturdy sons of toil who were doing whatever it was that they were doing in their corner of the library had started to whistle *"England's Coming Home"* in tuneless counterpoint to the sound of their power tools, we could still hardly hear each other.

Although we could hardly hear each other, it seemed that we were audible to passers by because an elderly gentleman with a venerable shock of white hair limped over to us and muttered to is in a pleasantly conspiratorial manner:

"So ye' r wanting to know about the history of Ormskirk are ye?"
We nodded our assent, and our new friend bent over towards us, and still in the manner of someone imparting a great and terrible secret at the risk of his own life, began to tell us exactly the same information that I had gleaned from the self-published History of Ormskirk a few minutes earlier.

Now, in all my cryptozoological travels both here and abroad, I have always wanted to find myself in a situation akin to that at the beginning of *An American Werewolf in London* where the two American students enter a pub on the remote Yorkshire Moors and the whole place comes to a veritable standstill. Feeling uncomfortable, they make to leave and one of the locals warns them to *"Keep on the roads"* and *"Stay off the Moors"*.

My dear friend and colleague Nick Redfern claims that when he was investigating the Berwyn Mountains UFO crash of 1974, he was visiting a small village in North Wales near to where the alleged crash supposedly took place when an old woman from the village warned him in a whiny Welsh accent:

"Don't go near the mountains my boy. You'll never be seen alive again.

61

Strange things happen up there"

before going into a complicated farrago about Men in Black, mysterious lights and animal mutilations.

Now, I love Nick like a brother, but I know that he has a tendency to hyperbole especially when it comes to trying to wind me up, or play inane practical jokes on me, so I take whatever he says in this vein *cum grano salis*. However, this morning in Burscough I finally found myself going through a very similar scenario, although in this particular instance the information which was being imparted to us by our mysterious informant wasn't the slightest bit sinister.

According to local legend in the first or second century AD, the flatlands between the Rivers Mersey and Ribble were mostly marshes and were dominated by Martin Mere - the largest lake in England. Around the lake were many small villages and the whole area was ruled by a Viking Chieftain called Orme (or Oarme depending on which source you read). He was known far and wide as a nasty and brutal piece of work, and it seems probable that like Vlad Tepes, known as "The Impaler" because of his jolly habit of dispatching his victims on a pointed stake, named himself after the dragon which was the fiercest beast in their folkloric pantheon. Vlad named himself Dracul or Dracula (meaning the Dragon) and it appears that Orm (whose real name was probably Kevin or Duane), probably did something similar.

Vlad Tepes is even today feted as one of the national heroes of Romania because most of his victims were the invading Turks, and so his crimes against humanity are overlooked because he did the Dark Ages equivalent of making the trains run on time. It seemed that a similar scenario can be applied to Orme. He converted to Christianity, built a church (kirk) on one of the few hills by the Mere, and founded a town called Ormskirk and everyone lived happily ever after.

We thanked our new friend, and having got completely fed up with the noise and bustle of this tawdry excuse for a library, we went back to the car and commenced our drive to Ormskirk. There, we had been told, was a much bigger library with far more extensive facilities including photocopiers, scanners and Internet access. There would also, we sincerely hoped, be no builders doing their inimitable if noise thing on the premises and everything in the garden would be relatively rosy.

62

Groovy, we thought, and drove west.

It was obvious that the whole country had once been marshland, and that much of it still was. Apart from the occasional stands of withys there were very few trees and the whole terrain was as flat as a pancake and criss crossed at intervals by interlocking waterways. There were patches of rushes and reeds, and even cotton grass (a species that I thought was confined to the south west of England) everywhere, and it did not take much imagination to envisage what this area would have been like in the years before the area had been semi-successfully drained. The ploughed fields revealed a rich fertile loam which had obviously been comprised of thousands of years of rotting peat and marshy detritus and the country-side felt as ancient and strange as anything that I had ever seen in England.

As we drove along, one of us - I cannot remember which - brought up the subject of F.W. Holliday and his bizarre theory of the Great Orms, which, he claimed, had once been found in every stretch of marshland across the United Kingdom, and indeed northern Europe.

Holliday believed the orm to be a gigantic invertebrate, a worm in the modern sense of the word. He believed its identity to be a giant, latter day form of the tullimonstrum a tiny prehistoric worm. Fossils of the tullimonstrum have been found in coal deposits near Chicago. Its name means Tully's monster, after its discoverer.

Holliday, otherwise an excellent writer, seems to have based this shaky theory, on the elongated shape of the tullimonstrum. The little animal resembled a spindle with diamond shaped fins towards the rear. Worms can reach amazing lengths. The ribbon worm can exceed 100 feet. But these soft bodied animals cannot support great girth or bulk. Even the ribbon worm is at most a centimetre across. The theory also ignores the obviously reptilian nature of the orm with its sharp teeth and scaly skin.

However, it is an indisputable fact that Holliday was right in one aspect of his theory. Most of the great marshlands of Europe have their own dragon legends, and it would be strange indeed if this area, once perhaps, the largest area of marshland in the Kingdom did not have its attendant draconian population.

We were convinced that we would eventually be able to find a set of leg-

63

ends to support this theory.

We drove into Ormskirk where we suffered our third body blow of the day. Although we found the public library almost immediately we couldn't find anywhere to park. We drove round and around Ormskirk getting more and more frustrated. We were determined to find somewhere to park and use the library, because we were convinced that somewhere within its interstices we would find some clue to the Orm of Ormskirk.

We continued to drive round and around, and then by dint of a particularly inane piece of signposting we ended up on a one way street which led us right out of the town, and before we knew it we were right out in the middle of the country. Now, I'm sure that if I were writing a psychic questing book (which I most definitely am not) I would claim that somehow the Dark Earl of Ormskirk had led us to the particular spot where we were to make our next major discovery, but it was nothing of the case. We were just driving round and around in circles getting hopelessly lost when we stumbled upon a small cluster of houses.
I am sure that if I were writing a psychic questing book (which I most definitely am not) I would claim that the aforesaid Dark Earl of Ormskirk, who would, of course, be the reincarnation of the original Viking warlord after which the town was named had not only led us to this desolate spot, but had made it absolutely impossible for us to find our way back to it. However, this too, would be nonsense. The truth is that we were completely lost and our road map was spectacularly inadequate for the purposes which we required it for. We were so lost that when, eventually we managed to find an ordnance survey map of the area, there were so many little hamlets of two or three houses marked on the map within a few miles of Ormskirk itself, that we had no means of knowing which of them was the one in which we had found ourselves, so except by the most spectacular happenstance, we are unlikely to be able to find it again.

It is a damn good job, therefore, that we had our trusty digital camera with us, because there, carved into the stonework over the front door of what seemed to be a very old house was a crest featuring a crown, a lion and a dragon! Finally we had some concrete evidence that there was a dragon associated with Ormskirk. Perhaps this was the Orm itself.

Although it was certainly many centuries younger than the time when the Norse warlord ruled the area this was unquestionably a dragon. During our travels around the country we have come across several places where dragon legends, and even dragon cults have been perpetuated throughout the ages.

The Vikings carried the image of the dragon on the prows of their long-boats. They went in fear of one particular sea orm known as the Shoney. According to the accomplished Fortean and North East historian Mike Hallowell this fear grew into a dragon worshipping cult that flourished along the north coast of England.

Beginning when the north was under Danelaw Viking ships would attempt to pacify the Shoney with human sacrifices. On each trip a human, bound hand and foot was tossed over board after having their throat slit in the hope that the Shoney would devour the unfortunate sacrificial victim instead of attacking the ship. There are records of the partially eaten bodies washing ashore up and down the North East coast and in particular around Lindisfarne or Holy Island.

This tradition seems to have persisted long after the Vikings had been converted to Christianity. Scandinavian sailors carried on the tradition of sacrifice to the sea dragon for centuries. Carved into the cliff face in Marsden Bay, South Shields is a pub known as Marsden Grotto. Victims of the Shoney cult were regularly washed ashore in the bay and were laid out in the pub's cellar for collection by the local constabulary. Amazingly the last of these sacrifices was discovered in 1928! It is amazing to think that human beings were being sacrificed to a sea dragon well into the 20th century.

Mike believes that a number of these Orms inhabited the net work of deep caves, both natural and man made, that stretch for miles beneath the area. These monsters were the basis for many local dragon legends. Mike himself claims to have seen the Shoney in the sea of Marsden Bay. A brown hump rising from the water much like the Loch Ness Monster in its "upturned boat" aspect. An elongate neck and tail were visible below.

If such a dragon cult could be active in the North East of England until recent times (there is even some evidence that it persists in some form to this very day), it is tempting to hypothesise that the same thing could have happened in the North West.

However, one swallow doesn't make a summer and one carved stone dragon doesn't make a cabal of sinister draconian cultists, but it is a nice idea.

We took a number of pictures, and then decided to give up our search for Ormskirk public library, and to instead drive straight to Southport where we knew that we would be able to find a major public library with books, a reference section, a photocopier and most importantly of all, a disabled parking place.

As we were approaching Southport, my mobile phone which has spent the entire journey ether in the breast pocket of my shirt or ensconced in a convenient nook on the dashboard of the Jag rang. It was Tim Matthews wondering where we were and what we were doing. We told him that we were about ten minutes away from Southport and made arrangements to visit him and his heavily pregnant wife Lynda when he had finished work at about five thirty.

We negotiated the maze-like streets of Southport like old hands, and parked in a disabled parking bay. By this time I was very tired. I have made no secret of my declining health, and over the past few days I had been doing far more than usual. It is my habit to have a siesta in the afternoon (which has always seemed to me to be an eminently civilised practise), and for the previous few days, not only had I not had my siesta, but I had driven in excess of 400 miles, done a reasonable amount of walking and far more typing and research than is usual in my life. I was really rather tired, and as it was a beautiful sunny day and the parking bay was in a quiet and relatively restful side street I decided to go to sleep for a while. Richard moseyed off to investigate the public library, and I took my camera bag, which made a perfect impromptu cushion and drifted up the figurative wooden hills to Bedfordshire.

Forty minutes or so later I was awoken by Richard. Tim had taken a late lunch break and was waiting for me in the reading room of the public library. With great reluctance I let the call of duty take priority over my desire to return to the arms of Morpheus and we locked up the car and strode off through the pretty covered walkways towards the public library.
The more time I spend in Southport the more I realise what a beautiful town it once was, and indeed in many ways still is. It has a faded *fin de siecle* feel about it, and its imposing Regency architecture still maintains a certain cultural veneer, in a pleasant contrast to the raddled old whore

that is Blackpool just across the River Ribble.

Tim was waiting for us outside the public library. He had been summoned back to work but he thrust a thick wodge of photocopies into my hand before he took his leave of us. I shoved them inside the zip-up compartment at the back of my laptop computer carrying case and promptly forgot about them as we went into the library in search of more information about the strange place in which we had found ourselves.

We soon found out that this particular part of the Lancashire coastline was completely unlike the rest of the county - and indeed the rest of the north of England. For one thing, it had a strange and unique wildlife.

Parts of the south coast of England, particularly Dorset, Hampshire and Kent are home to relict populations of some species of animal from mainland Europe that are at the very north of their geographical range. Animals like the Sand Lizard, the smooth snake and the natterjack toad are such European creatures that are not found in the rest of the country. However, it turned out that there are relict populations of both natterjack toads and sand lizards in the coastal areas on what would once have been the shores of Martin Mere. For some reason, this little part of Lancashire is like a tiny chunk of the south coast which has been transplanted three hundred miles north for no apparent reason.

The cryptofauna of the area is equally bizarre.

On pages five and six of a book called *Lancashire Magic and Mystery - Secrets of the Red Rose Country* (Sigma, 1998) Kenneth Fields writes:

"In January 1997, another alarming report came in about a strange black beast that had been seen on the Lancashire coast. The police in Southport received two separate calls within half an hour concerning a "black panther" seen at Ainsdale. Immediately a helicopter was scrambled together with an armed response team but the creature remained undetected.

When news of the sighting became public, many other people wrote in to the local newspaper relating similar experiences around the coastal dunes. One man told how he had met the creature three years before in Formby Woods and other witnesses related more recent encounters in the same area. Unexplained loud growling at night, a large black animal at least five foot long, and pets mysteriously killed were just a few of the reports.

However the beast of the dunes may have a supernatural explanation which is recorded in the folklore of this once isolated area. It seems that the ghost dog of Formby, known as Striker, has for centuries haunted the sandhills between Ainsdale and Formby. It is said to be a large, black hound with eyes which shine in the dark and its presence is surrounded by death"

There are also accounts of sightings of faeries and little people. However, unlike those in some other parts of the country, most such encounters reported in the lost marshlands between the rivers Ribble and Mersey are malevolent in nature. Fields reports:

"One traditional tale relates that in Penwortham Woods on the banks of the River Ribble a fairy funeral is sometimes seen. Anyone who is unlucky enough to view the strange spectacle should beware for it is an omen of imminent death. Two farmworkers were returning home at midnight, when by chance, they came across the cortege winding its way along a path that led from the ancient churchyard. In amazement one of the men gazed down upon the miniature coffin, then to his horror he saw that the face of the corpse was his own. Within a month the man was dead, having fallen from a haystack. Ironically his funeral passed along the same route taken by the fairy funeral".

There is an irresistible parallel between the Penwortham story of the Faerie Funeral and an old Devonshire folk tale, which I have collected for my ongoing book on Zooform Phenomena in Devon folklore, which I have been writing for at least twelve years, and which lurks on my computer's hard drive making me feel guilty whenever I think of it.. Theo Brown, usually considered the greatest Devonian folklorist of the 20th Century described, what is possibly the most famous piece of pure Black Dog folklore from the westcountry in *Tales from a Dartmoor Village* [1952]

"The Demon Huntsman. I have seen it stated that the whisht hounds emerge from Wistman's Wood, but I have never heard of them being seen by anyone. Baring-Gould, however, had a story from Moreton that seems slightly related. A moorman (according to one account, he lived at Chagford) was riding home one night from Widecombe (another account says

he lingered too long at the Saracen's Head at Two Bridges, but that would be somewhat out of his way) and passed a circle of standing stones. Suddenly and silently past him flew a pack of jet-black hounds followed by a dark huntsman. Recklessly the drunk moorman cried: 'Hey, hold on, mister; what sport? Give us some of your game!' 'Take that,' replied the huntsman coldly and tossed some-thing to him as he passed. The moorman caught at a bundle but could not in the dark distinguish what it was. He rode on home, dismounted, called for a lantern and peered at the object in the bundle. It was his own baby, dead and cold.

I have seen versions of this story from Normandy and Germany. The latter, however, concerns a Baron Rodenstein of Odenwald, who was thrown a piece of rotten flesh and subsequently lost his two best horses."

Although we could find no evidence whatsoever for the sort of Dragon Cult that Mike Hallowell had discovered in Tyneside we did discover evidence of both a dragon legend from the area and evidence of human sacrifices, although there is no evidence whatsoever that the two legends are in any way linked.

In prehistoric times there existed where the present day town of Runcorn now stands, a vast forest of mighty trees. According to the local legends this forest was haunted by strange and monstrous animals including a giant flying dragon who had his lair by the riverside near where Halton Castle later stood. This dragon was the scourge of the neighbourhood and laid the entire area of Cheshire and south west Lancashire to waste.

Deep in the marshes that surround Martin Mere was a sturdy blacksmith called Robert Byrch whose fame both as a blacksmith and as a manufacturer of suits of armour was known throughout the two counties. He had seen the dragon and had also heard it proclaimed in the local market places that a great reward had been offered to whomever could rid this part of the realm of the predations of this great and fearsome beast.

Robert noticed that although domestic livestock had been attacked all over the region, none of the animals that wandered in the vicinity of the forge itself has been attacked and he hypothesised that this was possibly because it was frightened by the fire from the forge. However, one November dusk the dragon swooped down and carried off a cow which had been grazing outside the forge, and although Robert didn't really care about his neighbour's cattle being taken, once the draconian predator was beginning to snack upon his own herds he began to get annoyed.

The angry blacksmith began to formulate a cunning plan to dispatch the dragon. He built a metal cow and covered it with the skin of a freshly slaughtered member of his herd with the head, horns, hooves and jaws still attached. He got inside the metal beast and in a scenario similar to the Trojan Horse he waited until the dragon carried him away, and then grasped a specially forged sword of toughened steel and cut his way out of the stomach of the bewildered dragon causing it to die in agony.

News of Robert's feat reached the ears of the King who was so impressed by Robert's achievement that he granted him as much land as could be enclosed by the skin of the dragon cut up into as small a set of strips as is possible. According to an excellent little book called *Lancashire's Ghosts and Legends* (from which I have cribbed much of this narrative):

"This is no fairy story or figment of the imagination. The aerial combat with the dragon could be seen years ago in graphic outlines on the wall over the kitchen fireplace at the Bold Arms hotel at Bold Heath. For years the inhabitants of Bold and Farnworth were familiar with the skin which hung above the bold family pew in Farnworthy church. There, high above the congregation, it hung for centuries until, in the 1870s the old relic fell to the ground. After much careful examination and the removal of the dust of centuries it was pronounced to be the untanned hide of a cow still bearing traces of the dragon's claws"

Richard, who, by the way, is responsible for writing much of the material on the subject of dragons and their kith and kin that is in this book, is in the process of writing what will, when it finally emerges, be the ultimate book on the subject. He has been kind enough not only to write some pieces for this present volume, but to allow me to quote directly from his *magnum opus* on the subject of strange dragon-like creatures seen in Southport:

"Ian Wharton told the author [Richard] of the strange encounters of two of his colleagues in the Parks Department working at Hesketh park, Southport. One man, Clive Everson approached Wharton one morning in having claimed to have just seen what looked like a Pterodactyl. A grey skinned, bat winged creature with a long beak and massive wingspan

had risen up out of the bushes in front of Everson and flown away leaving him dazed and alarmed.

A second man Percy Whaterton had seen two beasts answering the same description in the woods that backed onto his house. The were badly frightening the birds and Wharton thought they may be nesting in the forest. To date nothing further has been heard."

There are other similarities between West Lancashire and the dragon-haunted areas of Tyneside. As in the North West there is also evidence of a long tradition of human sacrifices in the area south of the River Ribble. On page 102 of his book Kenneth Fields notes:

"When the wanderings of the early nomadic tribes gave way to more permanent settlements, the nearness of a water supply was quite naturally of great importance. This dependence then gave way to superstition, the wells and rivers assuming mystical significance, becoming the abode of water gods who demanded homage. These waters were assuaged by casting in fruits, blossoms, animals and even unfortunate humans. Evidence of what is believed to have been a sacrifice to the River Ribble was discovered last (19th) century at Preston docks where a large number of both human and animal skulls were discovered. (...) The Red Moss mummified body of a girl from the first century may well have been a sacrificial virgin who was thrown into the marsh to appease the gods"

Probably the strangest thing that we discovered during that long afternoon was an alternative derivation of the name Lancashire, and one which, once again, has a very definite connection to monster killing. Again our main source is Kenneth Fields who wrote:

"There are romantics who question the widely accepted view that the name of Lancashire is derived from Lon Castrum which was the Roman fort on the banks of the river Lune. In their view Lancashire was originally Lancelot's Shire, having taken its name from the noble Sir Lancelot who ruled over part of the North West"

Sir Lancelot of the Lake, aka Sir Lancelot du Lac, or as T.H. White referred to him, *Le chevalier mal fet* (the ill made knight - a reference to his supposed ugliness), was, according to legend at least, the son of the Kong of France, who orphaned at an early age was adopted and brought up by a friendly and accommodating water nymph. Some sources claim

71

that this nymph was the same Nimue who entranced Merlin, and may or may not have been the Lady of the Lake, and other people have different ideas. It has been suggested that the Lake where Lancelot was brought up was situated in France, in Cornwall, in Wales and in the lost land of Lyonesse, but according to an authorless little book called *Lancashire Hot Pot - all you need to know about Lancashire* which was published in 1992:

"The Martin Mere Wildfowl Trust inland from the town of Southport opened with 260 acres in 1975, a fantastic resource for conservation and education with facilities for visitors and school parties who find plenty to marvel at all the year round. The shallow mere, once covering several square miles and dotted with marsh grit islands was reputedly the place where young Sir Lancelot grew up before joining Arthur's knights of the Round Table".

Although, to the best of our collective knowledge there are no legends alleging Lancelot as a slayer of dragons, he is alleged to have killed at least two giants (one of them in Lancashire) and a malevolent marsh troll.

As we left the museum that evening our heads were buzzing. What had originally seemed to be a perfectly simple zoological puzzle was rapidly becoming something far more complex. Whether or not the Monster of the Mere actually turned out to be a large fish as we still suspected, there was no doubt that it was inhabiting an area rich in folklore and accounts of high strangeness. It seemed that the marshy flatlands between the two rivers were a very strange place - a place, where to paraphrase Percy Fawcett talking about South America in the early 20th Century, "Nothing would surprise".

We drove around the marshlands for about an hour drinking in the sights, sounds and smells of the place in the light of our newly acquired knowledge. The salt marshes and swampy flatlands which had seemed strange and alienating before, now seemed like a magickal wonderland where monsters, Faerie folk and even Arthurian Knights could leap out at you from around the next bend in the road. This place had everything - everything that is except for a shop which sold Ordnance Survey maps.

It took us two whole days before we found an Ordnance Survey map of the area but we finally bought one, ironically at the same petrol station where we had originally stopped on our first approach to Southport. As we ate our evening meal, also purchased on my trusty Barclaycard at the same gas station I telephoned Tim Matthews for instructions as to how to get to his and Lynda's house. Much to my pleasure it turned out that they only lived just around the corner and so five minutes later we were driving up to their front gate where the whole Matthews Family; Tim, Lynda, Alexandra (aged two and a half) and Freya (due to be born the following Tuesday) were waiting to greet us.

We sat happily in the Matthews front room for an hour or so and drank tea and chatted desultorily. Tim and Nick Redfern are probably my two closest friends in the fortean community (outside my own household) and as always it was a pleasure to sit down and chat. There is always a feeling of *me casa es su casa* about our visits to either the Redfern abode or the Matthews household as, I hope there is when they visit us, and so we felt like we were in our own homes. As always we talked about everything apart from forteana, and we drank tea and played with Alexandra and the two dogs until it was time to drive back to Blackpool.

As the Jag pulled up outside Rob and Karen's home, so did a little red car being driven by Janet, one of the core members of LAPIS. The previous evening at the meeting in the pub we had spoken at some length about our plans for this expedition. One of the things that we had intended to do was to chart the decline and fall of Martin Mere itself. Our biggest theory was that the creature living in the lake was a wels catfish and that as these creatures are not native to the United Kingdom it would have to have been introduced from somewhere else.

We were hoping that the catfish would prove to be a particularly ancient one that had been introduced into one of the waterways in the area by Frank Buckland or one of his cronies from the Acclimatisation Society. However, as the mere had been essentially dried up until it was refilled (to a fraction of its former size) in the mid 1970s, we wanted to find out which of the myriad small ponds and waterways in the area had remained constant over the years since Buckland had visited the area.

Finally we now had all the relevant information at our fingertips, because not only had Tim Matthews photocopied a series of 1906 newspaper articles about the history of the mere for us, but Janet had spent an hour or so in Preston public library photocopying, not only a more modern account of the draining of the mere, but the relevant sections of Ordnance

Survey maps of the area published in 1894 and 1922. Together with the modern Ordnance Survey map we had purchased at the petrol station in Southport we now had all the information we needed at our fingertips.

We also had another ace up our sleeve.

When she was much younger Lynda Matthews had actually worked at Martin Mere and knew her way around all the complex interlocking waterways which surrounded the nature reserve. Although at the time of this visit she was so heavily pregnant that she could hardly move, by the time it came for us to do our follow up visit later in the summer, she would be far more mobile and would be an invaluable guide to us in our endeavours. We had also been promised help from a bunch of people from the LAPIS group and so we were confident that if, on the Monday, Pat Wisniewski gave us the permission to return later in the year in order to carry out our in depth investigations that we would have a pretty good chance of achieving our objectives.

The three of us were in a bullish mood, therefore as we strode cheerfully up the garden path to knock on the front door.. We were planning to sit down, have a cup of tea, and immediately start work on comparing the three different maps in our possession. However, it was not to be.

The door was answered by a short stocky fellow with a cropped haircut and a nasty looking scar on his temple.

Mr Downes, dear boy" he boomed at us in a voice delightfully reminiscent of that of the late great Keith Moon, *"How the Devil are you? Come in and have a drink"*.

It was Dave "Geordie Dave" Curtis, a childhood friend of Rob's who, in the short time that I had known him, had become a close personal friend of mine and Richard's. I had met him first at the 1999 LAPIS conference (he was the "Dave from Geordieland" alluded to in Andy Roberts's scurrilous article) and we had soon found each other to be irresistible soul mates with a keen interest in drinking, singing, causing mayhem and generally having a good time. Dave is the singer with a band called *Happy the Man* and plays a mean guitar.

In certain circles I am known as "The Wild Man of Cryptozoology", an appellation which, even I admit, is slightly deserved), and I suppose that if I am Cryptozoology's Keith Moon, then Dave is forteana's Viv Stan-shall. We have appeared together, guitars and bottles of booze in hand, at UFO and fortean conferences around the country, where we sing, shout, get drunk and behave in what we, at least, feel to be an uproariously funny manner. Whether anyone else feels the same is a moot point but on the whole people seem to enjoy our antics and continue to let us get away with it. We have had many an entertaining lost weekend together and as soon as I saw his smiling face greeting us at the door I knew that all hope of doing any work that evening had gone completely out the window.

I was right!

Saturday 1st June 2002

Dave is like a cat. He always makes a great fuss when you meet him, and he greets you effusively. When he feels that it is time to go, however, he just puts his metaphorical tail in the air and walks off into the sunset. It was no great surprise, therefore, when the next day, we had a brief telephone call from him saying that he had enjoyed himself but was now going back to his family in Co. Durham. The previous evening he had been drunkenly fantasising about spending the whole weekend in Blackpool and joining us on our research trips, but the next day he obviously had thought better of it.

Although I am desperately fond of him, I am not going to pretend to him or to anyone else that I was 100% disappointed. I love him like a brother, but I know full well that any protracted length of time that we spend together (especially when I have a couple of credit cards in my pocket would soon turn into a drunken lost weekend of the most debauched variety, and neither my bank balance, my health, or indeed this investigation could handle that at this point in the proceedings.

We were, in fact, at a stage of slight impasse in our investigations. There was nothing much that we could do now, until we had conducted our interview with Pat Winsniewski on Monday. We knew that we would have to set aside a day to examine the documents that Tim and Janet had photocopied for us, but we also knew that the following day, the weather forecast was poor, and predicted thunderstorms, whereas today the weather was as glorious an English summer day as one could hope to

77

have.

We decided to spend the day doing a little bit of quasi fortean sightseeing. However, at least partly because of our drinking session on the previous evening we didn't actually get out of bed until lunchtime, and it was well after two before we found ourselves driving the Jag round Blackpool's diabolical one way system in search of a car park near to the UFO Museum.

As recounted in my book *Only Fools and Goatsuckers* (2001), I once heartily offended a very right on subcontracted employee of Channel 4 by describing a little village in the heart of the Mexican desert as *"The biggest shithole that I have ever visited"*. I can now take that back. I have now visited Blackpool, and the peasant village with the open sewer, the cockfighting, the child prostitutes and the dead dogs beside the verge of the road is now only the second biggest shithole that I have ever visited.

Rob Whitehead, who has lived here for something like seven years says that its twin towns should be Sodom and Gomorrah and that it should be napalmed. Despite the fact that I do, believe it or not, always do my best to find something pleasant to say about everywhere that I visit, I have to agree with him. Blackpool, or at least the parts of it that I visited, made Las Vegas look wholesome.

Just like I had done a few days before in Southport, I wondered about the history of the place.

The history of Blackpool, it turns out, mirrors the history of the area of Lancashire which we had been exploring quite closely. The Fylde was an area of forests and bogs dating back to Roman times which was inhabited by people known as the "water dwellers". The Romans built a road which went through Preston, and then continued west to a port situated north of Fleetwood.

The ancient parish of Bispham was recorded as far back as the Doomsday Book. In 1416 a district in the Fylde, including an area known as "Le Poole" the "pool", was a stream which drained the area known as Marton Mere into the sea near Manchester Square. The stream went through peat bog land, which turned the water to a black colour, hence the name Black Poole.

Marton Mere still exists, although like the place that is very nearly its namesake on the opposite side of the River Ribble it is now only a tiny fraction of its former size. So far we have not been able to establish why the two places have such similar place names. There must be a link, but what is it? If anyone reads this book and knows the answer we would love to find out.

In 1602 people began to build cobble and clay huts near to the area of the "Pool". In the Bispham parish register, the names of " de Poole" and "de blackpoole" were mentioned, and the name of "Black Poole", was short-ened to the present day name of "Blackpool"
As I have already discussed in my book *The Rising of the Moon* (1999), as a direct result of the ill health of His Majesty George III, and his pre-scribed cure of seawater, sea bathing became popular amongst the aris-tocracy and gentry, and although such people had begun to visit Black-pool in the 1720's (and the first guest house was opened in 1735) by 1780 the seaside resort now had four hotels and four ale houses in Black-pool, and two more in the district of Layton.

Blackpool soon began to develop into a major seaside resort. The devel-opment was so dramatic that Blackpool's population rose from 473 in 1750 to 47,348 by the turn of the century. The railways reached Black-pool in 1840, and allowed travellers from the mill towns of Lancashire and Yorkshire to have cheap excursion trips. And how they came in their thousands to Blackpool, the top entertainment centre of the North.

- The North Pier was built in 1863
- Central Pier 1868
- South Pier 1894
- The Tower 1894
- The Grand Theatre: and the Big Wheel at the Winter Gardens in 1896.

In 1879 Blackpool was the very FIRST place in the world to introduce the new concept of electric street lighting, which was empowered by the electric arc street lighting system. The first Mayor of Blackpool, Dr. Wil-liam Henry Cocker, was elected when the new town was made a Bor-ough on the 21st January 1876. The famous Blackpool Illuminations were created in 1912. With the Illuminations Blackpool now had a longer season than any of its rival towns.

As discerning readers may have guessed I culled most of the above information from a selection of webpages about Lancashire local history which told me more than I could ever need to know about the history of Blackpool. What none of them could explain, however, was quite how, what had started off as a genteel seaside resort - a Northern Brighton, if you will, had turned into such a god-almighty cesspit.

Richard and I gazed out of the window in horror as we cruised round and round the filthy streets looking for somewhere to park. *"This is everything that's wrong with Britain today lumped together, multiplied by ten and put in the same place"* growled Richard as another gang of drunken louts with skinhead haircuts, ear-rings and tattoos, false plastic breasts and cans of lager lurched across the road in front of us, as their girlfriends staggered on behind. He told me how, in the late 1970s, his grandfather had described seeing groups of small children playing in the sea at Blackpool while lumps of human faeces floated around them.

Family groups of the most horrific looking people were everywhere. The very people who you see on every television programme about the dangers of paedophilia, marching up and down clutching badly spelled placards and threatening to *"torch"* the *"gaff"* of someone they believe to be a *"nonce"* just because he has a passing resemblance to someone whose features appeared in a badly printed photograph in *The News of the World* were there with their children. Their little girls wore the sort of revealing clothes that as a parent myself, I would shudder at an eighteen year old wearing.

I am no feminist, but I try to treat everyone with respect regardless of age, sex, colour or creed. However, I would argue that for a woman of any age to be wearing skin-tight pink silk trousers and a tiny pink crop top, made of a sheer material which not only plainly revealed her erect nipples, but was obviously designed to do so, and emblazoned with a motto reading "I'm a sexy bitch" is inappropriate. It is demeaning not only to herself but to all members of her sex. When the person wearing it is no older than nine years old it beggars belief. When her sister – older , but still obviously under the age of consent – is wearing an equally revealing outfit and a T Shirt proclaiming her to be a "Porn Star" one really does feel that one is losing grip on reality.

I feel like little Bobby Zimmerman writing *"A Hard Rain's Gonna Fall"* as I try and give you just a few of the impressions we gathered of this

80

disgusting cess pit of a town.

* We saw women in late middle age wearing enormous cowboy hats covered with red glitter, pink fairy wings, and T Shirts reading, Kiss me, spank me, F*** me.

[AUTHOR'S NOTE: The word F*** is what was printed on the T Shirt. As regular readers of my work will know, I am quite prepared to use the old Anglo-Saxon word `fuck` where it is appropriate. To use an F and three asterisks is merely vulgar]

- We saw a group of drunken lads in their late teens or early twenties staggering about in the gutter. One was dressed as a circus ringmaster, one was dressed as a clown, one as a gorilla and one as the devil

- We saw gangs of drunks with the cross of St George painted on their faces, standing in a line as they urinated in the gutter as their jeering girlfriends stood by watching

- We saw a joke shop which sold sex aids and children's toys side by side in the window.

The streets were full of rubbish, the sea looked like an open sewer, the gutters were full of a river of piss and vomit in which MacDonalds wrappers floated merrily, and the air stank of grease, cigarette smoke, sweat and decay.

In the middle of this was the Exhibition of the Universe (formerly known as the Blackpool UFO museum).

We finally managed to find somewhere to park the car and we walked unsteadily towards the museum itself. Out of nowhere came two Gypsy women who accosted us. One, smiling thrust a piece of "lucky white heather" bound up with cotton and silver foil into my pocket. She thrust two little glass pebbles into my hand and told me that they would bring me luck.

I think that I can be forgiven for my first thought which was that these two women saw us as soft touches and wanted us to give them money. I protested that I had no money (which was true), and I turned out my pockets to reveal that all I had was a wallet full of cards, some fluff and a couple of receipts. She smiled, kissed me on the cheek and told me that this was going to be a particularly important year in my life, and that on the 25th or 26th July my life was going to come to a very important turning point. As will be seen later in this book, she was quite right.

I thanked her, and she smiled again and said *"God Bless You for your kindness my dear"*. I apologised again for not being able to give her any money, and she smiled back and said that she just wanted me to have her good luck charm. *"The Lord loves those who love the Lord"* I said. She smiled again and disappeared into the crowd.

I felt an incredible sense of kinship with the two Gypsy women. In many ways I am a very conservative person. I own my own house, I drive a Jaguar, I vote Conservative, I am an unashamed Royalist, I went to a public school, my brother is a Vicar, both my father and my brother have been decorated by the Queen and I try to pay my bills on time. In other ways, though, like the gypsy women I am very much an outsider and feel very detached from mainstream society. My feelings of alienation were multiplied a hundredfold whilst I was in Blackpool, just because it is a particularly horrible place, but even in my own home town of Exeter I feel detached from society as a whole. Like Brian Wilson once wrote *"I just wasn't made for these times"* and I feel that my belief systems, my system of values, and even what I do for a living are so far removed from the way that most people spend their lives, that I have little or nothing in common with the vast majority of people in society. I like to think that the gypsy woman did indeed have the second sight attributed to people of her race and saw in me a kindred spirit; someone who through happenstance and accident of birth, but also through a series of lifestyle and belief choices had become alienated from the other people around him. I like to think that when she saw those traits in me as I limped towards her leaning on my walking stick, she reached out a hand of comfort and friendship to me. I hope that my response touched her as well.

Although we don't often talk about it, I am fairly sure that Richard who was walking slowly by my side as we ambled through Sodom, and Graham back at the CFZ in Exeter feel equally at odds with mainstream society. The fact that one has adopted the mantle of Gothic flamboyance and the other is (despite the fact that I love him dearly) one of the most peculiar people that I know, and that they have both decided to throw their lot in with me suggests that I am right in this assumption.

We walked slowly down the promenade towards the entrance to the museum. It is situated below the Sea Life Centre where sharks in enormous tanks circle endlessly within their plastic universe to amuse the proletariat. I went up to the young fellow on the entrance booth and asked to speak to David Boyle. He smiled at me, muttered into a walkie talkie, and within minutes David bounded out beaming to meet us.

We had met briefly at the 2001 LAPIS conference and although we had got on OK, I had got the impression that he was less than impressed with my cynical attitude towards the subject of UFOlogy. I got the impression that he was even less impressed by the showing of my ridiculous art movie *The Owlman and Others* which featured Gay cowboys, a pre-op transsexual nazi on a bicycle, naked lesbian sex and a total absence of any coherent plot. My mentor within forteana has always been the surrealchemist wizard of the western world Tony "Doc" Shiels, and it is what I have learned from him, together with what I have gleaned from the books of folk like John Keel that have spurred me on my way. I came to the conclusion many years ago that the Omniverse is fundamentally stupid, and though the folk who claim that there is a Universal consciousness, and an over-riding intelligence behind it all may well be correct, no-one ever claimed that the over-riding intelligence behind the Omniverse had to be sane! Indeed all the available evidence points to my theory that there may well be a Galactic Overmind, but if there is he or she (probably she) is as mad as a spoon.

My books, films and records, have mostly been, either like this one, straight objective representations of what I have seen on my travels, or joyous celebrations of the fundamental absurdity of everything. Some people call what I do art, some people just find it silly, and a few, I am afraid are offended by it.

I was horribly afraid that David Boyle, a cheery chubby little man in late middle age might well have been one of the latter.

However he greeted us heartily and proceeded to show us around his museum. My first impression was amazement that one man could have achieved so much in such a relatively short period of time. The museum has only been in its present location for two years, and three years at its previous home, but it had a sheer volume of exhibits and information to see that would shame many professionally run museums dedicated to less esoteric subjects. What made this achievement all the more extraordinary was that it had all been put together by one man.

In the ten years since I founded the Centre For Fortean Zoology we have published nearly twenty books and twenty eight issues of our journal, as well as maintaining a web site and several one off exhibitions. We have mounted two major overseas expeditions and a number of less exciting ones within the UK, and on the whole I am proud of what an organisation consisting of three core members, half a dozen assistants, consultants and volunteers and an ever changing number of hangers on have achieved in the last decade. David Boyle's achievements, however, in half that time, put us completely to shame.
There is only one problem.

I didn't understand more than ten percent of what David was trying to tell us during our whistlestop tour of the museum, and what I did understand, seemed to me at least, to be a farrago of esoteric nonsense. David cited such texts as *The Wizard of Oz* and *Dr Who* as being of massive spiritual importance. He claimed that *Star Wars* was a mirror of our past and *Star Trek* was a mirror of our future, and that he was sure that the spaceships of the Galactic Federation were going to make themselves visible to the people of the Earth within days as they were bound to want to avert the threatened nuclear war between India and Pakistan.

He went into great deal of detail about chakras, angels, crashed UFOs, The International Monetary Fund and the Freemasons (BAD) and the Council of Ascended Masters (GOOD). He pointed out a poster on the wall which seemed to imply that The Count of Saint Germain, who might have been a spiritual master who had lived for hundreds of years, might have been an aspect of the Wandering Jew, but was most likely an eighteenth century con-man (and probably several different eighteenth century con-men), was higher on the ranks of the ascended masters than was Jesus Christ. He went into a long explanation about how the logo of MasterCard and the symbol for one particularly malevolent group of extra terrestrials were identical. And what did that mean? He kept on asking us.

He showed us a magnificent model of a chupacabra - a beast about which I know quite a lot, and which, in my book *Only Fools and Goatsuckers*, I have theorised is the Latino manifestation of a worldwide genre of vampiric entities given physical form by the religious and sociopolitical hopes and fears of the people who live in the communities in which its predations take place. David believes, however, that it is a pet of the "Space People". I groaned inwardly.

When he told Richard - a life long Dr Who fan - that the BBC had cancelled the show not because of dropping audience figures, bad casting and even worse story lines, but because it was educating too many people about the real meaning of the universe I thought he would choke. Much to his credit he didn't. I wanted to tell him that I had once spent an entire evening in Nevada trying to seduce the ex wife of the writer of *Battlestar Galactica* on a golf course until we were interrupted by automatic sprinklers but I thought better of it.

He tried to explain to us in tortuous detail how the layout of the stones at Avebury in Wiltshire was a precise mirror of the layout of certain geological features of the plain of Cydonia on Mars. He claimed that the Avebury stones were a symbolic representation of the vagina and womb of Gaia the earth mother, and how certain hills and mounds across Salisbury Plain were exact representations of her breasts and belly. He said something about the "face on Mars" which I didn't even begin to understand, and then as the *dénouement* he showed us a model of the Galactic Federation Council meeting around a Round Table, obviously based on a cross between Michaelangelo's *Last Supper* and King Arthur's court as portrayed by Walt Disney.

When our tour came to an end, neither Richard or I knew what to say. Neither of us were qualified to comment on his highly complex n dimensional mathematical extrapolations, but the bits that we *did* understand just seemed to be arrant nonsense. Even the Council of Elders were obviously 1970s shop window mannequins in silly costumes, and the model of Jesus looked more like one of the Bee Gees. However David was, and is such a terribly nice man and obviously 100% sincere. To criticise him would be nothing short of churlish - to quote Harper Lee it would be *"as bad as shooting a mockingbird"*. This was the only man in downtown Blackpool who was doing something because he believed in it. He wasn't there to fleece the throngs of unpleasant proles out of their money, he just wanted to save the Universe in the best way that he could. Who the hell are Richard and I to try and stop him?

We asked him why he had chosen to put his museum in Blackpool. He gave us two completely different answers. First he said that he had chosen Blackpool, because, surrounded as he was by all the tackiness and stupidity of the modern Babylon, that the Government (and presumably the powers *behind* the Government) would leave him alone. About half an hour later I asked the same question and he said (possibly more truthfully) that it was because he wanted to reach the general public rather than just the cognoscenti. Unfortunately it appears that the general public just ignore him whilst the cognoscenti laugh at him. This just ain't fair for such an obviously sincere man even if he *does* claim that rabbits are genetically engineered bio computers.

David Boyle is one of the nicest men I have met in a long time, and if only for his undoubted sincerity he deserves our wholehearted support. Just don't ask me to explain what the hell he was talking about. And by anyone's standards he is the only showman in Blackpool with any integrity whatsoever, so shine on David.

After leaving the museum we decided to take a drive into the countryside. We had drunk our fill of the fleshpots of Babylon and needed to see some fields and trees for a change.

We had nought but the most basic idea of where we were going to. We just wanted to drive around the countryside for a bit. We took the first available exit off the motorway and headed in a vaguely easterly direction. We didn't really care where we went as long as it was in the opposite direction to Blackpool, and there were no gangs of yobs wearing plastic comedy breasts urinating by the sides of the road.

As we drove along we noticed quite how different the countryside north of the River Ribble was, to that of the area that we had been investigating. Although it was pretty enough, it lacked the strange, slightly menacing ambience of the flatlands between the Ribble and the Mersey. During our periods spent in the Southport library the day before we had noted the same thing when it came to folklore. Lancashire is full of folklore - there are plenty of stories about ghostly headless horsemen, mysterious witches and the like, but it was only in the flatlands that the folk tales were as macabre as the ones that I recounted above. It was almost as if, elsewhere in the county the folk-tales were just that; stories to amuse and entertain children and adults of rural communities, but in the flatlands they took on a much nastier and much grimmer nature. I think it was then that we started to think of the dunes and marshes that we were beginning to know quite well as "The Lancashire that Lancashire forgot".

As we drove along, I suddenly noticed a familiar name on a roadsign. *"Hey, lets go to Longridge"* I said, and without bothering to explain why to Richard, I turned suddenly to the right, and we headed towards another place of mild cryptozoological interest. According to one, almost certainly apocryphal report cited by Kenneth Fields, the town got its name during the Civil War when Cromwell, leading his army up the tortuous hills which led to the town said *"What a long ridge this is"* but even Fields admits that this story is almost certainly nonsense. What is far more interesting, to us at least, is that it is allegedly one of the places where one of Britain's most mysterious unknown animals - The Dun Cow - was said to have lived and died.

Longridge is only one of at least three places in Lancashire where the Dun Cow, a Faerie beast sent to Lancashire by the Gods, was an enormous mythical bovid which wandered freely over the moors and allowed herself to be milked by anyone who came. However many milked her, their pails were always filled, until a malicious witch milked the cow all day into a sieve, and the animal, exhausted by the efforts, died. It is said that there is a house in the town which has one of the ribs of the beast displayed over its doorway. However, they say the same thing about the other three sites in Lancashire where the beast was meant to roam, and so as Longridge turned out to be a fairly dull and unprepossessing township with nothing much to recommend it (and as we were getting hungry) we made our way slowly back to Blackpool, where I fulfilled a telephonic promise of mine and brought Richard and Rob a sumptuous Chinese meal.

On the way back Richard told me some of the other legends about the Dun Cow which have been recorded from other parts of the country.

Another Lancashire legend of this fabulous creature takes place at the top of Parlick Hill between Chipping and Bleasdale in North Lancashire, where even today there are several sunken tracks which terminate at Nick's Chair on Blindhurst Fell. According to this particular variant of the legend, the Dun Cow roamed the Bleasdale, Chipping and Browsholme terrain, giving an abundance of milk to all and when thirsty it drank from a well called Nick's Water Pot on the top of Parlick.

The top of Parlick is not a place where one would expect to find a well, hence the reference to, *'the top of Parlick'*, in this legend, was surely due to the significance of Parlick to the culture that evolved around the

Bleasdale Circle from the time it was built at some time between 1900 B. C. and 1720 B.C.

If so, this would suggest the Dun Cow legend may have had Neolithic Origins. With Nick's Water Pot, reminding one that, Nick's Chair, still existed on the top of Blindhurst Fell, the ridge that extends from the top of Parlick to the top of Fairsnape Fell. (Fairsnape Fell is where the mid-summer sun would rise when seen from the Bleasdale Circle on the day of the summer solstice around 1800 B.C.).

Dun Cow references are rare in the old Celtic stories. The two best known are those of Manannan mac Lir, the pagan Celtic god of the sea, after whom the Isle of Man is named. These were; *'cows with twisted horns, a speckled cow and a dun cow that were always in milk'*. The other Dun Cows were those of the Irish pagan Celtic god Lugh, a sky god whose counterpart was Lleu Skilful Hand in the old Welsh legendary stories. The Irish legend tells how Lugh advised the people of Ireland to satisfy the demands of an aggressive character called, Bres, who required as a tribute; *'vast quantities of milk from cows all of the same colour'*. The advice given by Lugh was that they should create; *'magic cows from bog stuff so they were all Dun Cows'*.

Guy of Warwick, one of King Arthur's Knights is also said to have encountered this mythic cryptobovid. The story is much the same as the others – the benign beast is giving milk for free to the starving peasantry when a wicked witch makes her run out of milk by milking her through a sieve. However in this version, the Dun Cow is understandably annoyed and ran amok causing death and destruction wherever she went. Along came good Sir Guy who dispatched her before going on to destroy other creatures of cryptozoological interest across the country including a huge green dragon in the grounds of Longwhitton Hall, in Northumberland.

After a pleasant evening in the Chinese restaurant we returned to Rob and Karen's via an off licence, where, once again, my trusty credit card was called into action to purchase whisky (for me), cider (for Richard), and Bicardi (for Rob and Karen), and then we went back to their abode to drink. During our journey back to Rob's we shared reminiscences about the more ridiculous aspects of our UFOlogical careers. Richard and I told Rob about the events chronicled in my 2000 book *The Blackdown Mystery*, and Rob managed to reduce both Richard and me to fits of the giggles by telling the story of a particular video purporting to show a fifteen second snippet of an unconvincing alien face appearing over the back of a sofa in the middle of a three hour video tape featuring, what Rob described as *"a bunch of fat hicks, showing off for the video camera by sitting around throwing pillows at each other"*.

Apart from our mutual interest in forteana and Cryptozoology, I think that the main reason that Richard and I work so well together as a team is that we both have very similar senses of humour. We both appreciate the absurd and ridiculous and are quite happy to talk childish drivel to each other for hours until we both collapse into helpless laughter. I used to be like that with my Mother when I was younger. We could retreat into a surreal nonsense world as well, but it was something we lost as I grew older and we grew apart. I have had glimpses of that surreal nonsense world again during my adult life, but it has only really been since Richard, Graham and I have become fortean zoology's version of Biggles, Algy and Ginger, that I, too now dwell in a highly ludicrous version of Arcady. *Et in arcadia ludicouso ego.*

Still laughing, we arrived at Rob's house.

Karen was watching *Top of the Pops.* Although I have been a rock music aficionado for three decades now, and have an ever expanding collection of compact discs (which, with the advent of file-sharing software, threatens to take over the house as I download as many obscure albums from the net as I buy from second hand shops), I haven't watched *Top of the Pops* in years. Watching it, I began to feel middle aged.

Like the fleshpots of Blackpool earlier, the content of the programme alternately annoyed and bored me. Most of the anodyne pap merely passed me by as I drank and chatted happily (whilst secretly wishing for a complete set of *Led Zeppelin* CDs in my brief case). Then on came Billy Bragg – a troubadour whose work I had admired whilst he and I were railing against the greater excesses of Thatcherism in the 1980s. Together we marched (figuratively at least) in support of the miners, against Cruise missiles, and in abhorrence of Ronald Reagan and the Star Wars programme. In many ways his music had been a soundtrack to my life between the ages of twenty four and thirty. This time, however, I felt as alienated from him as I had done from any of the scumbags that I had been unfortunate to meet in Blackpool that morning.

He was singing a dismal dirge called *"Take down the Union Jack"* which, apart from a vaguely anti-jubilee ethos didn't actually say anything very concrete. The most appalling thing about the whole performance wasn't that the song wasn't particularly good or that the Big Nosed Bard from Barking actually had nothing of any interest to say, but that neither the TOTP presenters or the audience seemed to pay any attention to what he was saying.

Twenty five years before *The Sex Pistols* had released the notorious *God Save the Queen* – a single which had been re-released to overwhelming apathy that week. When it first came out I, like thousands of others bought it and played it clandestinely. It was banned by the BBC. Possession of it was forbidden at my Boarding School and the record was earnestly discussed in Parliament. Twenty five years later even John Lydon had mellowed:

"Let me remind you what being British is all about. This is our country, this is our flag, they're our monarchy. They don't work too well at the moment but let's make the fuckers do a good job. Let's get rid of the useless ones and keep a few of the goodies."

Like me he was appalled at the vacuousness of Bragg's dismal record. *God Save the Queen* had been an adolescent scream of anger against a stultifying society ruled by a tedious old man called James Callaghan, where the only alternative seemed to be the appalling prospect of nineteen years of Thatcherism. To John Lydon and to me Billy Bragg was just a self-congratulatory middle aged git whose desire for a hit record was far stronger than any real desire to attack the things that are *really* wrong with this country – many of which have been caused directly by the people whose policies he has supported all his life.

"Why didn't anybody boo the bastard?" growled Richard as he stumped off to bed. There was no answer that I could give, but later as I drifted off to sleep I made a silent bet with myself that Bragg would end up on one of Tony Blair's "people's" honours lists before too long,. I mentally added Bragg to my list of people who would be first up against the wall when my particular revolution came, rolled over and went to sleep.

90

Sunday 2nd June 2002

I woke up in the middle of the night in agony. The sound of distant thunder filled the skies, it was lashing down with rain and my head and chest felt like someone was performing a trephination and open heart surgery using a Black and Decker Workmate, but without the benefit of anaesthetic. I have been prone to migraines in thundery weather for many years but this was one of the most extreme ones that I had ever experienced. For some reason I also had violent muscle cramps causing excruciating back, neck and chest pains, my lungs were full of fluid and I truly thought that I was going to die. Death has no real fears for me - I like to think that there is some kind of life hereafter and if I'm wrong there is just oblivion - but the manner of my passing scares me rigid. Over the years too many of my loved ones have died in painful, undignified and even agonising circumstances for this not to be the case. I know full well that I shall die someday . I'm also honest enough with myself to realise that my poor health would suggest that this may be sooner rather tan later and that I am unlikely to make old bones, nevertheless I had no intention of dying at four in the morning in somebody else's spare bedroom in Blackpool.

Although I could hardly move I managed to prop myself up in bed. I reached across to my briefcase which as well as my ongoing notes, address book and various other bits and bobs, contained my impressive pharmacy for the trip. I grabbed a handful of mixed pain killers, decongestants and tranquillisers and washed it down with a mouthful of Jack

Daniels. Muttering a quick prayer to St Elvis (the Patron Saint of Poly-pharmacy) that I would survive this criminally stupid act of prophylaxis (which from experience, I knew was the only course of action that would bring even temporary relief), and longer prayers to the various deities who battle for supremacy in my psyche since my estrangement from Mother Church, and I tried to get some sleep.

Surprisingly sleep refused to come, and the horrific mixture of chemicals and booze that I had ingested had little or no affect. I was still in pain, I still couldn't breathe properly and I wasn't even slightly drunk. Even the pharmaceutical opiates I had swallowed seemed to have no effect what-soever. So for two or three hours I sat up in bed listening to the rain and praying for some kind of oblivion which never came.

The sheets of lightning split the sky apart. As the drugs began to numb the edges of my consciousness, although I could see the room lit up by flashes of white/purple electric heat, I couldn't hear the thunder. The only sound that I could hear was Richard's rhythmic snoring. Still sleep wouldn't come, and for the first time in ages I was praying for a stronger panacea. I wanted so badly to drink from the waters of Lethe, but all I had was the remains of a half bottle of Jack Daniels. I had already taken well over the stated dose of painkillers and far more valium than was sensible. All I could do was drink, and as I swallowed the rest of the bourbon, neat, n three swallows, I could feel each mouthful hit my empty stomach as if it were sulphuric acid.

My heart began to thump, irregularly, and it was as if I could hear every single beat. My back and my rib cage started to hurt and I could feel a tightness across my chest and a dull ache down my left arm. Not for the first time in my life I was convinced that I was dying.

One of the worst things, for me at least, about being on the road during an investigation under these conditions is that I have nowhere to which I can retreat. As I write this a couple of weeks after my return to Exeter after the first leg of the expedition, I have my own particular environ-ment almost entirely sorted. It is one in the morning, I am sitting up on one elbow in my own bed. My cat, Helios 7, is asleep on my right hip. *Seinfeld* is on the tv, the pharmaceutical opiates are beginning to lull me into a warm cottonwool womb, I am surrounded by my books and CDs and I feel comfortable. I can watch what I want, listen to what I want, read what I want and in the middle of it get on with writing this book.

Unfortunately I spent too much of my life alternating between states of psychotic isolation, and a drug induced haze. Intermittently over the past ten or fifteen years I have left my womb like isolation, and in a brief burst of energetic enthusiasm I have carried out lecture tours, concert tour, collecting trips and expeditions, and then I have retreated back into my comfortable womb-like bedroom to write about what I have done.

There are times when I am at home, in my study/sitting room or in my bedroom that I feel as ill as I did that night in Blackpool. At home, however, I have ways of dealing with my problems, even if it is only lying in the dark chanting along to the Radha Krishna Temple on my CD player, or hugging the cat. Although, even at home when I feel as ill as I did I can feel close to death - since my mother died and my health has deteriorated I have been very aware of my own mortality - it is somehow easier to deal with when on is at home in one's own space. As I write this I am comforted to know that I am surrounded by my own things, my daughter is asleep in the attic room above me, and the teddy bear I had as a child is sitting next to the teddy bear my mother had in the 1920s in the wardrobe at the end of my bed.

As the first rays of dawn began to pierce the slate grey Lancashire sky, I managed to prop myself up into a position that was just about comfortable enough for me to drift into an uneasy sleep. As I drifted off I wondered, seriously, how much longer I am going to be able to do this. My past has caught up with me. I spent so much of my life abusing my body in a desperate attempt to combat the demons which, even now, inhabit my tortured psyche, that my health has been irreparably affected. The irony is that during the past six years from 1996 during which I have made serious and concerted efforts to deal with my mental health problems, although I have learned skills to help me deal with my psychosis my physical condition has deteriorated to such a degree that I seriously wonder whether I shall ever recover.

It seems bitterly unfair, that now, when for the first time in my life I am able to travel around the country for any length of time without being either drugged into semi consciousness or being accompanied by a retinue of minders, my physical condition was deteriorating to such an extent that I don't know how much longer I shall be able to do things like this.

"Damn it" I thought. "I'm going to carry on doing what I do until I can't

do it anymore, and I'll face that problem when I get to it".

Despite myself I began to smile, and as I finally managed to drift to sleep, I started to think positively about the expedition so far and what we had learned about our elusive quarry.

When I finally woke up, Richard had disappeared and Rob and Karen had gone out for the day. I was feeling decidedly shaky and I had a splitting headache but at least the pain in my back had gone. I started working on this manuscript, and I had only been writing for about fifteen minutes when Richard strode in cheerfully clutching my credit cards and a plastic bag containing a nourishing breakfast of the haddock, chips and red wine variety. Richard had decided (quite correctly) that I would be quite happy for him to shove a nourishing breakfast on my credit card, and the fact that he would have to forge my signature in order to shove hot food and red wine down my throat would be a small price to pay. I have always been insistent that my two closest associates - Graham and Richard - know how to forge my signature as and when it is necessary. I love them and trust them like brothers. They take care of me in ways that no-one else ever has, and together we have had, and I hope will continue to have adventures beyond anything that most middle aged men aspire to.

Things were beginning to look up, and as I noshed my fish and chips and swigged red wine out of a pint glass, Richard and I began to discuss the next stages of the investigation. We decided that we had to have a concerted look at the photocopies that both Tim and Janet had provided us with and do our best to learn a bit more about the history of Martin Mere. However, first of all there was a bit of donkey work to do and some more of the previous day's notes to type up.

I still had a diabolical headache and so, feeling relaxed for the first time in about fifteen hours I went back to bed as Richard typed up some of the of the notes he had got from the library a few days earlier.

This time I had no problem at all in going to sleep. The red wine and the high protein made me feel ridiculously replete, and I happily fell into a deep sleep. I opened the window and the soft air of an afternoon in early summer was more soporific than any of the drugs that I had taken the previous night. My sleep was punctuated by the smell of grilled sausages and loud cheers - a family several doors down was obviously having a

World Cup party. I remembered the last time I had been on the road during the world cup.

During the summer of 1990 my ex-wife Alison and I travelled around the country with a classic 1970s rock band, *Steve Harley and Cockney Rebel,* (for whom we were working at the time), and a touring party of loveable eccentrics that we seemed to have a knack of picking up along the way. The whole adventure was chronicled in my book *Road Dreams,* and was one of the happiest four weeks of my life. It was probably the happiest oasis in an otherwise difficult marriage, and I drifted off to sleep remembering happy days in an endless summer whose memory is still very dear to me.

What seemed like minutes, but was actually something in the region of three hours later Richard woke me up with a cup of tea. I was feeling happy, comfortable and contented. I was also ready to start some serious work. By this time it was beginning to rain quite heavily. We had been invited to a skywatch by Janet and the rest of the LAPIS posse, but bearing in mind the weather, the amount of work we still had to do, and the fact that we had one hell of a day ahead of us on Monday, we telephoned Janet, made our excuses and returned to work.

The first thing that became apparent was that the history of Martin Mere was far from simple. Once upon a time it had been the largest lake in England, and it had been drained essentially in order to reclaim the rich, fertile flat lands of the lake bottom for agricultural use. Tim had got us a long article on the draining of the Mere which had originally been published in the *Southport Visitor* in 1906. It was written by a Rev Bulpit, and described a vast lake, five miles long and two and a half miles wide. (The entire article is reproduced in Appendix Two). As we had discovered, the Mere was a strange and ancient place, which had once been renowned as the haunt of dragons, giants and spectral black dogs. As we were to discover later in our investigations, these were not the only Zooform creatures to be reported from the dark peaty waters, and we wondered whether the draining of the Mere might not have had a deeper meaning than mere agriculture (no pun intended).

Could the draining of the Mere not be symptomatic of something deeper and far more primal - the struggle that we have seen over and over again in our expeditions between so-called "civilised" man and his pagan and wild past? Was the destruction of the lake part of a wider suppression of the ancient and arcane in favour of the tameable, the mundane and the easy to understand? The more we looked at the history of the place the more complicated the whole investigation became.

Janet had provided us with some unreferenced photocopies from a guide-book to local history. It gave a brief history of the place, far more succinct than the overblown Victorian verbiage of the Rev W.T. Bulpit:

"Until the end of the seventeenth century Martin Mere drained into the Douglas at its eastern end and Rufford gets its name from the causeway crossing, now followed by the A59 main road. In winter the Douglas ran into Martin Mere, increasing its area, but in summer the flow was in the opposite direction and clear evidence of flowing water is to be found in soil sections in Mere Sands Wood. In 1692 Thomas Fleetwood of Bank Hall, Brethren, devised a plan td drain the mere from the western end direct into the sea; in order to do this some mechanism was necessary to prevent the sea from entering Martin Mere. Fleetwood leased the mere from the other landowners for 'three lives and thirty one years' and built sluice gates at Crassness at the end of a one and a half mile long, 24 feet wide, channel cut along the length of the mere by two thousand men. The mere itself lies 10 feet. lower than the high-water mark of spring tides and the sluice gates closed when the sea water rose above the level of the canal. However, sand and mud eventually blocked the gates and the mere again flooded. In 1714 the gates were raised 20 inches but whilst this prevented blockages it affected the drainage and the mere remained largely flooded."

Something that was becoming increasingly obvious was that whereas the main body of the lake had vanished by the beginning of the 19th Century, there were isolated bodies of water, and patches of marshland which, according to the 1892 and the 1928 maps in our possession, as well as the contemporary O/S map that we had purchased in Southport which had remained inviolate. There were also several cuts and sluices which criss crossed the landscape which essentially had not changed for nearly two hundred years.

Lynda Matthews had told us about these whilst we were back in Exeter and the investigation was still in its planning stages. As a teenager she had worked at Martin Mere, and she promised that as soon as she were mobile, she would help us trace the winding capillaries of the marshland which although the lake was now only a fraction of its former glory still criss crossed its ancient boundaries.

It seemed also, that even without the lake, the fertile marshland had supported a rich and varied zoofauna throughout the 19th and 20th centuries:

*"The Ribble Marshes and its surrounding mosslands have been a haven
for waders and wildfowl from time immemorial, and every now and then,
when the pumping station at Grossens has failed to cope with heavy
rainfall, Martin Mere has returned to something of its former glory as
birds quickly found the flooded fields. In the 1950s, excessive flooding in
1954, 1956 and 1957 resulted in action being taken as a result of the
strong agricultural lobby, and on 21 July 1961 a new pumping station
was opened at Grossens."*

It also seemed that the network of small streams, drainage ditches,
marshes and ponds that we had noted as remaining inviolate despite the
draining of the lake itself, as well as providing a haven for wildlife, had
been prone to flooding on a regular basis and that it was only the advent
of the Grossens Pumping Station, technically advanced for its time, that
finally drained the remains of the Mere, only forty years before our arrival:

*"Jointly operated by the River Grossens Drainage Board and the
Crossens River Board, this was capable of removing 450 million gallons
of water a day. Pumping is now adjusted to maintain a level in the
ditches such that the surface soil can drain, but the complete drying out
of the peat is avoided in order to prevent soil shrinkage. With this pumping capacity the mere had apparently disappeared for all time and virtu-
ally the whole bed of the ancient lake was turned over to agriculture."*

It seems ironic that only a decade later moves were afoot to start filling
the Mere up again. This time due to the vision of the legendary naturalist
Sir Peter Scott:

*"A glance at the 1974 Ordnance Survey map still shows areas of 'marsh'
close to Mere Brow and Tarlscough, where at Holcrofi's Farm in 1970
there were still large areas of rough summer-grazed pasture that was too
wet to be ploughed. This area of 363 acres was acquired by the Wildfowl
Trust in 1972 through the vision of Sir Peter Scott who had known the
region well from his visits to watch, catch and ring part of the wintering*

Pink-footed Goose population. Initially it was the geese that were the main reason for establishing a reserve on Martin Mere, but who was to guess what would develop? In an early brochure Sir Peter wrote 'We have the opportunity to make a great step forward for conservation... The mere is an 'oasis' of primitive and unspoilt 'wetland' which can be turned into one of the finest natural nature reserves and wildfowl refuges in Britain'."

Work started on the Wildfowl and Wetlands Trust reserve in 1972 and it was open to the public three years later. By then, Sir Peter's vision of a unique haven for wildfowl in the North of England had been realised, and an astounding new environment - which though artificial, had all the attributes of the original lake, (although being a fraction of the size) had been created:

"Geese have been coming to the south Lancashire mosses at least since records began and, at the time of the establishment of the Wildfowl Trust Reserve at Martin Mere, some 14,000 visited the area every winter. Immediately after the war the flocks. could be counted in hundreds rather than thousands and numbers have gradually risen over the years. The majority of the south Lancashire Pinkfeet are of Icelandic origin; about 70% of the total breed in the Thjorsarver valley in central Iceland. Of the rest, a small number of birds are from Greenland but there are no records of Spitzbergen Pinkfeet which winter mainly on the continent of Europe. On southerly migration the birds often stop at sites in Scotland and for some years the presence of two Ross's geese identified the Lancashire flocks on their journey south. It is not unusual to find other species of geese with the Pinkfeet and a few Greylag, Whitefronts, Brent and Barnacle Geese occur every winter; Snow Geese, too, can sometimes be found and Lesser Whitefronts have been recorded. Bean Geese are more frequent than was at one time thought and there is at least one record of an Eastern Greylag shot from a flock of Pinkfeet at the same time as Bean Geese.

Geese can always be found on the mosslands in winter and seen during the day from roads between Altcar in the south and Rufford in the north. Since the early part of the century the birds have fed to a large extent on waste potatoes left on the surface, and there can be little doubt that the mechanical harvesting of potatoes has had much to do with the increase in numbers. In this they do a useful job in removing decaying tubers. For the rest of the time they feed on barley or wheat stubble or graze on the saltmarshes of the Ribble. Occasionally they turn to carrots that have been left in the ground before harvest and cause damage, but this can be avoided to some degree by siting carrot fields near roads. Another agricultural problem is 'paddling', where large numbers of geese trample the ground, so preventing aeration of the crop roots and proper drainage.

In the past, geese mainly roosted on the Horse Bank, on the mudflats of the Ribble off Southport, but now they spend the night inland in several places on the mosses, including Martin Mere. The maximum winter count of 36,000 birds was made at the end of January 1982, when over a third of the world's population of Pinkfeet was present in southwest Lancashire, 13,000 of them at Martin Mere. Usually wild birds can be seen from at least one of the hides on the mere. The very fine collection of wildfowl now maintained at the Reserve is one reason why many wild birds drop in. Perhaps the most significant visitors in recent years have been the Whooper and Bewick's Swans which, until some ten years ago, were rare visitors to the Ribble. Now, flocks in excess of 200 can be seen on the saltmarshes of the Ribble Estuary National Nature Reserve; many of these birds visit Martin Mere, particularly at night and during hard weather."

When we had started our quest we had been under the impression that all that we were looking for was a large fish. This might well still be the case, but as we sat quietly at Karen's dining room table looking at the sheer volume of information that we had been able to glean in just a few short days it was hard not to imagine, or at least discuss the possibility, that we were on the track of something far stranger. We were beginning to run up against a figurative brick wall however. We needed Lynda to show us around the complex network of waterways and drainage ditches and we needed permission to investigate the Mere itself. The first would have to wait until after Lynda had given birth to her second daughter, and there was nothing that we could do about the first until we had conducted a face to face meeting with Pat Wisniewski on Monday.

Assuming that we would be allowed to carry out the investigation in the way that we wanted we would have to return later in the summer with a larger contingent from the CFZ back in Exeter. We had been promised the assistance of the LAPIS posse as well as Tim and Lynda, and possibly some of their friends and colleagues. There would be three, and maybe four of us coming up from the westcountry and so we would have no shortage of manpower. However what were we going to do with them? After the best part of a week on the road, were we any the nearer to discovering what it was that we were actually looking for.

During my personal history of investigations into quasi fortean phenomena I have become fascinated with the concept of Zooform phenomena. When I started my chequered career as a cryptozoologist I desperately wanted to be able to explain everything that encountered within the terms of reference laid down by Darwin, Linnaeus, and Mendel, within a universe defined by conventional Euclidean geometry and four relatively safe dimensions. I soon found that this dream of mine was completely impossible.

Most mystery animals can be placed into one of three categories:

a) CRYPTIDS: Animals whose existence has not yet been accepted by mainstream science, and which are either completely unknown or are thought to be either locally, nationally, or globally extinct.

b) PSEUDOCRYPTIDS: Out of place animals. Animals of known species which are just found in the wrong place at the wrong time, usually (but not always) because of deliberate or accidental introduction by human beings.

I soon discovered that some mystery "animals" actually appear to be no such thing. They are not animals at all and are not even animate in the strictest sense of the word. In 1990 I coined a name for this group of "Things" (as the late great Ivan T Sanderson would have described them), and dubbed them ZOOFORM PHENOMENA - a term which has since been accepted by the fortean establishment and has come into common usage.

The term ZOOFORM PHENOMENA has been greatly misunderstood over the years. I have been accused of claiming that all "mystery animals" which cannot be *proved* to be flesh and blood, carbon based lifeforms are some sort of psychic beings with an objective reality of their own. Whilst I am quite prepared to believe that all the living things on earth cannot necessarily be gauged using a purely zoological model or set of references, I have never claimed that all Zooform Phenomena have any sort of objective reality. Indeed, I am convinced that precisely the opposite is the case. Zooform phenomena are things which *appear* to be animals - this includes hallucinations, misidentifications of natural phenomena, inanimate objects which appear to be animate and a host of other categories which do, I admit, include some non-physical beings with an objective reality.

Because of my own history of mental health problems, and indeed sub-stance abuse, I have made a particular study of the links between a wide range of quasi-fortean phenomena and mental illness. I have also investi-gated the links between such things and the use of psychoactive alkaloids both in a recreational and a ritual manner, and I have also drawn links between zoofortean phenomena and political/religious belief patterns most notably in the troubled areas of Central America where both civil war and the predations of the chupacabra are rife.

Could any of these factors be at work here? We had already established the link between the marshes and the original lake and a whole gamut of strange phenomena. We had discovered stories about monsters and mys-tery animals and archaeological evidence suggesting that until relatively recently human sacrifices had been carried out to placate the marsh spir-its. Over the years the CFZ and its associates had investigated similar things across the UK, and indeed the world. There was evidence of ritu-alistic behaviour involving Morgawr the Cornish Sea Dragon, The Shony of Tyneside, the Chupacabra and many other Zooform phenom-ena besides. Could something of a similar ilk be going on here?

On the other hand, if what we were looking for *was* merely a giant fish we had evidence linking the Southport area with the very man who had introduced wels catfish into several other parts of the country. We had also shown that there were quite a few ponds, ditches and streams which (on paper at least) looked as if they were perfectly capable of supporting a medium sized wels catfish for many years. We had also shown that the area had been flooded so many times that although the mere itself was now only fed by water pipes and artificial means, there would be no dif-ficulty in a catfish from another pond or waterway being washed into the mere during one of the periodic floods.

As far as we could see, on that rainy Sunday night in Blackpool, the jury was still completely out on the subject and there was nothing really that we could do until we had a considerable amount more information. We had reached a standstill, and there was really only one thing left for us to do. We went down to the pub.

Monday 3rd June 2002

In a total and dramatic contrast to the way that we had started the previous day, we woke up bright and early and ready to start the next stage of our investigation. After all, in many ways this was going to be the most important day of the trip. This was the day that the investigation so far had been leading towards, and more importantly it was a day, which if it didn't go well, would bring our adventure to an abrupt halt. Everything hinged on our lunchtime meeting with Pat Wisniewski. But whatever happened at least we were going to get the hell out of Blackpool.

Karen and Rob were still asleep as we decamped. I wrote them a brief thank you note as Richard loaded the car up with our worldly goods, and then we made our move. A few days before we had bought a street plan of Blackpool, and we used it to make our exit as quickly as we could. As the twin town of Sodom and Gomorra disappeared over the horizon behind us we started to plan the day's event.

It's probably something to do with being brought up as an expat but since I was a small child I have always found myself feeling at home in an area after I had only been there for a few days. It is, I think, a measure of quite how badly I despised Blackpool, that despite Rob and Karen's hospitality, I felt no empathy whatsoever with the place and will be quite happy if I never have to go back there again in my life. However, once we had crossed the River Ribble, and driven through Preston and were on our way to the flatlands again, I found myself feeling surprisingly nostalgic. Despite the fact that this we had only been exploring the area for a few days, I was becoming quite fond of the place and it would be a wrench - albeit a minor one - to leave. I remember reading something similar in one of Gerald Durrell's books about his adventures as an animal collector in South America. I had certainly become fond of my little sector of San Juan on the island of Puerto Rico, and of the part of Mexico City where we had established our base back in the early months of 1998, but I could never, in my wildest dreams of imagined that I would start to feel the same way about a fairly anonymous part of western Lancashire. However, I was beginning to be fond of the place and as the Jag sped through the flattened grey-green landscape on the way to the Mere itself, I felt a tiny pang of wistfulness at the thought that this would be the last time that I would be visiting the place for a while.

Then my mobile phone, on the dashboard, rang. Richard answered it, and started to laugh sympathetically. It was Lynda Matthews. She and Tim had intended to meet us at the mere for coffee, that morning and we had been planning to talk through the next stage of the project with them after our meeting with Pat Wisniewski. However, it was not to be. Lynda had gone into labour and with the best will in the world was not going to be able to do anything that day except for to have her second baby.

I pulled into a convenient lay-by and Richard passed me the telephone. Much to my amazement Lynda was actually *apologising* to us for having (as she felt it anyway) 'let us down'. I got quite cross with her, and told her not to be so silly. There are far more important things in life than chasing for giant fish, and having a baby is one of them. For someone in the early stages of labour to be even thinking about doing anything else except for sitting back with her feet up and waiting for the midwife is completely ridiculous, and I told her that.

I could hear her smiling on the other end of the telephone. She promised to let us know when the baby was born and then rang off. I slipped the automatic gear box into 'drive' again, and we were just about to resume our journey when the telephone rang again. It was Tim. This time *he* was being apologetic. I told him much the same as I had told his wife a few moments earlier, but he insisted on apologising profusely for having stood us up. Again I told him not to be ridiculous and he laughed, promising to let us know when the baby was born, and rang off.

As I told Richard before resuming our journey, I really am extraordinarily fond of Tim and Lynda. He is, sadly, one of the most notorious men in contemporary forteana, and to my mind at least, his sordid reputation is ill deserved.

Before the fell out over something or other a few years ago Andy Roberts, commenting about the hate campaign being waged against Tim by a left wing activist called Larry O'Hara, wrote:

"I can't comment on Tim's political past. It interests me but I wouldn't believe anything he or O'Hairy told me about it and corroborative evidence is thin on the ground. O'Hardup is a master of the smoke screen, he piles 'fact' on 'fact', tarts it up with a bit of supposition, some high handed moral indignation and expects us to believe it. Hey, Larchy - we're ufologists y'know.

To believe what exactly?

Well that's the problem - O'Hara rants and he raves but rarely does he actually make any points. Bottom line in this case is, I think, this:

- *O'Hardly believes Tim has been both far left and far right. One of these is good - because that's the gang that Larry is in.*

- *O'Hardly believes that Tim is not motivated by ideals but only by his 'paymasters' - which in this case Lairy believes to be 'the government', or to put it in strict UFOlogical parlance, 'them'.*

Furthermore O'Hardof hearing believes that Tim has been sent among us to cause havoc, get us to believe in bad things, to do this and that and to generally be disruptive.

Sorry, but I don't buy it."

Neither do I. The whole affair is highly dubious from the start and any claims that the powers that be could actually be *bothered* to send someone - even Tim - to destabilise UFOlogy is a measure of quite how paranoid the world has become at the end of the 20th Century and the beginning of the 21st Century. Andy continues:

"Personally I think Tim would act how he does whether he was involved in extreme politics, model train building or inter-species full contact La-

crosse. He can be a bastard and he rubs people up the wrong way and sometimes his lack of tact can be frightening to the head-in-the-sand brigade. But I've never had a problem with him and quite a lot of other ufologists manage to get along quite nicely in Tim land."

I have never described myself as a UFOlogist, except during the strapped for cash years of 1997/8 when I was desperate for the money, but I, too have always managed to get along quite nicely in Timland.

But I have even more respect for Lynda. She is *truly* a remarkable woman. My love life over the past quarter century has been spectacularly unsuccessful probably because women like Lynda Matthews are thin on the ground in the modern world. Tim is probably just as difficult a man as I am to live with, but Lynda not only has stood by her man with a tenacity and good humour that would put Tammy Wynette to same, but is still resolutely her own woman. She is a devoted wife and mother, and I am sure that if I had met someone like her twenty years ago I would not be the curmudgeonly old bachelor that I have seen myself become. However, if that *had* happened, my life would have taken a completely different turn and I don't suppose you would be reading this book now - probably because I wouldn't have written it! As my long departed Grandmother used to say *"If ifs and ans were pots and pans - we'd all be travelling tinkers"* and there's absolutely no point in wondering what might have been if my life had taken a completely different course.

We suddenly realised that it was getting on for lunchtime and we hadn't actually eaten anything that morning and we were both getting rather peckish. Finding food in rural Lancashire on the Golden Jubilee Bank Holiday was easier said than done. Everywhere was closed. Unfortunately for the intrepid travellers on the track of giant fishes, the 21st Century equivalent of the explorer's trading post - the roadside gas station complete with supplies of pies, sandwiches and various types of deli stuff - didn't seem to exist. All the service stations in this particular part of Lancashire seemed to be depressingly unimaginative and sell nothing but petrol and motor supplies. However after about half an hour we managed to find a gas station that sold food as well, and so I parked the Jag, topped up with petrol and went into the service station to stock up on supplies. By this time, both Richard and I were ravenously hungry and so we quickly filled our shopping basket with a wide variety of goodies which would have easily sufficed to have taken care of both of our missing breakfasts and lunches (and leave a little extra provender to fill up any chinks that might arise later in the afternoon). We ambled up to the checkout and although I was mildly disconcerted to find that we had spent over twenty quid on groceries, there was no other realistic option to I handed the girl behind the counter my credit card, and waited for the transaction to take place.

Whilst the young lady in question was fiddling with the cash register in a gloriously inept manner I leaned on my walking stick and gazed out the window trying to identify a small yellow and brown bird which was hopping around on the far side of the road. I was jerked out of my reverie by the adenoidal voice of the young woman behind the counter who sounded particularly pleased with herself as she proudly informed me that the transaction had been refused. Suddenly my good mood vanished. *"There must be some mistake"*, I stammered, trying to hide my confusion and also fighting a difficult battle against losing my temper, as the check out girl started to bluster at me:

"Someone's going to have to pay for this you know....." she proclaimed triumphantly.

I told her that I was perfectly aware of this and suggested as politely as I could that she try and telephone the authorisation number on the back of the card. This she did - twice - but the office at Barclaycard seemed to be perpetually engaged.. Was there, I asked, a cash machine in the area?

"No, it's broken" she said gleefully, and reiterated that someone was going to have to pay the bill.

I got my emergency credit card out of my wallet and gave it to her. She muttered something about not wanting to accept it as I had already been refused credit on one card, but reluctantly she took the other card and the transaction went through successfully. I breathed a silent sigh of relief. My emergency card had been near its limit and I had been a little worried that there was not enough credit left on it (unlike the one which had been refused which I could have sworn had at least five hundred quid on it).

Feeling the checkout girl's eyes boring into the back of my head as we went back to the car, I reached for my telephone. As soon as I was in the car I telephoned the Barclaycard Customer Service department, and much to my surprise (considering that the checkout girl had apparently found it impossible to contact them) I got through first time. Somewhat to my dismay, I was forwarded to the Fraud Department. This was a little disconcerting but my conscience was clear and I was hoping that it would be easy to sort the matter out.

As it turned out it was. I had only been issued with the credit card a week or so before, and as I had immediately driven up to the north of England - more specifically, to Blackpool which, it turns out is the fraud hotspot of the United Kingdom. The good folks at Barclaycard just wanted to check that my card was in my wallet where it was supposed to be and not in the possession of some geezer called Methadone Pete. Despite the fact that the affair had caused me some little embarrassment at the petrol station, one has to admire their perspicacity. So, happy again, and secure in the knowledge that I did after all have half a grand's worth of credit left on my card, and even more secure in the knowledge that we had a bag full of expensive and delectable groceries in the passenger footwell we drove off to find somewhere where we could eat our lunch before we arrived at Martin Mere for our meeting with Pat Wisniewski.

We found a lay-by overlooking a particularly attractive stretch of the Liverpool-Leeds Canal, and as we watched a family of coots dithering busily across the surface of the water we munched away on our breakfast. The Exeter Health Authority community dietician who is masterminding my ongoing attempts to lose weight would probably be appalled at the idea of her patient happily scoffing chargrilled chicken and polony sausage sandwiches for breakfast, but boy did they taste good?

As we ate, we pondered upon the importance of the canal to the local environment, and also its possible relevance to our quest for the monster of the Mere. The completion of the Liverpool line of the Leeds and Liverpool Canal in the late eighteenth century saw the development of Burscough Bridge into the most important canal town in Lancashire. Burscough became a bustling transport centre, it was a staging post for the packet boats that carried passengers between Liverpool and Wigan, some of whom would transfer to the stage coaches travelling along the turnpike road to Preston and the North.

The traffic on the canal continued to grow in the nineteenth century, it was heavy and varied. Boats carried coal from the Lancashire coalfields through Burscough on the way to the Liverpool docks and brought commodities for the fledgling industries that sprang up around the canal, such as imported grain for Ainscough's Flour Mill. Manure was brought from the dray horses and middens of Liverpool and dropped off at the muck quays along the canal, then used to fertilise the reclaimed farm-lands of South West Lancashire and further improve the area's agricul-tural output.

Burscough was home to the provender stores which delivered proven and hay to the stables along the canal. Unlike many of the other waterways dug during the great canal craze of the 18th Century, the Leeds-Liverpool canal continued in economic importance well into the 20th Century. Even the coming of the railways in the mid-nineteenth century did not have an immediate catastrophic effect on the economics of west-ern Lancashire - the Leeds and Liverpool Canal, the longest and most diverse canal in Britain, was still carrying nearly 2.5 million tons of cargo in 1906. The coming of the railways only boosted the importance of Burscough Bridge, sited as it was at the junction of two mainline rail-ways which only served to encourage manufacturing industry to locate in the area.

Still a focal point of transport routes, Burscough continued to thrive. For many of the passenger trains travelling between Liverpool and Scotland, Burscough Junction and Preston were the only stopping places before reaching Carlisle. In 1914 it was reported that over 100 trains a day were calling at Burscough Junction Station.

By the time we arrived at the beginning of the 21st Century, however, the heyday of the town was over, but its past left a heritage of a compli-cated interlocking waterways. As we finished our lunch and drove off towards Martin Mere for our meeting with Pat Wisniewski we saw for the first time how these little waterways lead off the main canal. They are all obviously part of a drainage system, but just by looking out of the car window as we drove past it was impossible to see whether they were overflow ducts to stop the main canal from bursting its banks, or whether they were sluices meant to drain the surrounding reclaimed agricultural land into the canal.

The more we drove through what had once been the bed of the greatest lake in England, we could see that the entire area was still criss-crossed with these tiny waterways and although it was no longer a lake *per se* it was still an area of wetlands to rival the Norfolk Broads or Romney Marsh.

Before we knew it we had arrived at the Wildfowl Trust Reserve. Suddenly we began to feel ridiculously nervous. An awful lot hinged on this meeting. After all the cock ups of the day so far, we were beginning to feel like nothing was going to go right. Knowing our luck something was bound to go wrong, we agreed pessimistically. Probably Pat had been called away and the pivotal meeting of the investigation would not take place.

We sat in the car park for a few minutes composing ourselves, before taking a deep breath and walking, intrepidly towards the main doors of the Visitor's Centre. For the first time that day everything started to go right! We walked smartly up to the glass fronted entrance booth and an employee of the Nature Reserve whom we had spoken to in passing on our previous visit came bounding up to meet us. He had a broad grin of welcome on his face.

"You two are the guys who are coming to see Pat about catching the big fish ain't you?" he enthused happily. Before we had a chance to reply he was off again, *"yeah, we've all been looking forward to you guys coming. It's gonna be great".* He ushered us in, told us Pat was on his way down to see us, and then disappeared, all without giving us a chance to do more than grunt a few syllables in reply.

I looked at Richard happily. If everyone was going to be as friendly and as enthusiastic as this then suddenly the whole project looked like it was going to be back on track again. Before we had a chance to say anything, however, Pat Wisniewski came down the open plan wooden staircase in the middle of the entrance lobby to meet us.

Now this is a disgusting generalisation, which if I were to use it in any other context would certainly be condemned as being sexist or racist or something or otherist but there are two types of male birdwatcher. There is the retired Colonel with bifocals, who looks at ornithology with a slightly liverish and patrician attitude, and there are the geezers who look or at least behave like Bill Oddie. Pat Wisniewski is most definitely of the second persuasion, and anyone of a certain age who can remember watching *The Goodies* on Wednesday evenings during the 1970s will know exactly what I mean.

Within minutes of meeting Mr Wisniewski we realised that we had definitely fallen on our feet.

110

During the course of our conversation over the next hour, which took place in his wood panelled office which overlooked the vista of the nature reserve, it turned out that Pat was not only sympathetic to our cause, but that he was a devotee of the world of Bernard Heuvelmans - the 'Father of Cryptozoology' who died in 2001, and had been the honorary consulting editor of our journal *Animals & Men* for nearly seven years. He was also (like Richard and me), a keeper of exotic reptiles, and he even came from Richard's home-town of Nuneaton - a rather depressing little market town in the Midlands. By this time we were on Cloud 9. If Pat had told me that his favourite singer was Scott Walker and his favourite whisky was *Jack* Daniels it wouldn't have surprised me, we had so much in common. Could it get any better?

As it happened, it could.

Whereas our conversations with Bernie a few days before had led us to believe that there was a certain level of doubt surrounding whether or not there was a mystery creature actually *attacking* swans on the Mere, just a few minutes conversation with Pat left us in no doubt. He, for one was convinced that there was something there. Yes, he told us, what Bernie had told us was essentially right. There was at least one bird which had been on the lake the previous winter which had suffered from a specific neurological disorder. He suggested that if we wanted to know more about the details of this disease then we should contact the Veterinary Officer at the WWT Headquarters at Slimbridge in Gloucestershire. However, it seemed that this was *not* necessarily the bird that had been attacked.

It was almost certain that the original report of a bird being attacked was, indeed, as Bernie had claimed, this brain damaged fowl suffering some kind of seizure.

The description that Bernie had given us on our previous visit, and indeed the symptoms described by Pat were very reminiscent of a bird suffering from terminal lead poisoning. Until changes in regulations requiring replacement of shot made from lead, by non-toxic alternatives, it had been estimated that over one million wildfowl in North America alone, died each year from picking up lead shot, mistaking it for grain or for the gravel which they use to grind up their food. Waterfowl that are active bottom feeders are most likely to be poisoned by ingesting or taking lead shot into their gizzards. G.B. Grinnel published observations of waterfowl poisoning by lead shot ingestion as early as 1894, in the wildlife magazine, *Forest and Stream.* Other countries have also reported poisoning from lead shot. An investigation of swan deaths in England revealed that half may have been from lead-poisoning as determined from the correlation of blood-lead to the amount of lead shot in their gizzards.

Chronic lead poisoning is associated with starvation and weight loss. Before death, birds may lose up to 60% of their body weight. On the other hand, acute lead poisoning in ducks, swans, and geese may leave them so weak as to be unable to fly, even before much loss in body weight has occurred. Observations in the early 1900s, in which remains of lead pellets were found in birds' gizzards, described waterfowl having a rattling in the throat, with the bill held open much of the time and dribbling a yellowish fluid. Often the birds couldn't fly or walk because of progressive wing and leg muscle paralysis. On land the tips of the primaries dragged the ground and on water the wings floated loosely on the surface. One study in the 1950s found that after the fall and early winter hunting season, 12 percent of ducks had gizzards containing lead. Ingested lead pellets usually disappear from the gizzard in about 20 days, either by eroding or passing through the digestive tract. But even if the bird doesn't die during that initial period, the chronic effects of poisoning may last.

Waterfowl are also vulnerable to lead-poisoning from lead fishing sinkers or split shot, the slotted weights fisherman use on their lines. Between 1984 and 1990, 17% of common loons, found dead or dying, collected in Minnesota were shown to have been killed by lead-poisoning. Fishing sinkers can be easily lost or discarded into the environment where they may end up in bottom sediments of lakes, ponds, or other water bodies, or along shorelines, piers, embankments, or rock jetties. If an angler's hook or line were to get tangled in weeds or other obstructions, the sinkers may be lost in the water where they could easily be ingested by water birds feeding on seeds or other food. Sinkers may also be dropped or discarded on land close by the water bodies; here they also could easily be picked up by waterfowl. In shallow water areas discarded or lost fishing sinkers are extremely persistent and thus may be available to waterbirds for hundreds of years. Natural deposition and sedimentation processes may eventually cover the discarded sinkers; but activities such as boating or dredging may disturb sediments and uncover the lead sinkers again. Receding water levels due to drought, tidal effects, natural subsidence, or to intentional drawdowns in reservoirs also could make sinkers readily available again.

It is tempting to hypothesise that if the affected swan *was* actually suffering from lead poisoning rather than from an undisclosed neurological disorder, this is actually supportive evidence for the presence of a large fish in the Mere. A large bottom feeding fish such as a wels could well have disturbed the bottom sediment of the lake and possibly unearthed the remains of a carton of shotgun cartridges which may have been buried or dropped before the lake was refilled in the early 1970s.

Gerald Durrell described a similar series of incidents at Jersey Zoo, when an unusually dry summer had exposed mudflats that had been covered by water since the German Occupation during the second world war. Durrell hypothesised in *Menagerie Manor* that a civilian living in Jersey, buried his shotgun cartridges rather than hand them over to the occupying Germans (and be accused of collaboration), or hide them in his home (and risk being shot by the SS on suspicion of being a member of the Resistance).

However, as Pat explained, the brain damaged swan did not explain the series of incidents that had occurred during the previous February. Swans had been attacked on at least two other occasions, and on at least one, although the bird had struggled free and made it to the shore it was obviously in pain because something had damaged its wing. The eyewitness reported seeing the bird struggling as if something was trying to drag it under the surface of the water.

Obviously the three incidents had been lumped together - probably by the journalist who quoted Pat as saying that the fish-like creature he had seen some four years before was "the size of a car". *"It was obviously nothing of the sort"* he smiled, and gestured towards the sofa that Richard was sitting on. *"It was about that big - or maybe a little bigger"* he laughed. *"The trouble is that the story came out in the newspapers the weekend before half term, and I'm sure that the people who covered the story considered that we were doing some sort of a publicity stunt to get more visitors over the holiday...they were never going to take the story seriously".*

This explained the hyperbole. The sofa that Pat had compared the creature he had seen to was about six and a half feet long. We might not be looking at a wels as big as the sixteen footer that was caught in the Danube during the nineteenth century, but we were still talking about a fish far bigger than any other of this species ever caught in the United Kingdom. Pat, too, thought that it was probably a wels. *"They have been caught around here"* he told us. Apparently one of the angling clubs in Southport had caught a couple of small wels in recent years in a pond in the district. There was no doubt, therefore, that there were *some* wels in the district. But could one have grown to a length of seven or eight feet in Martin Mere?

Pat believes that it could. After all there are certainly quite large fish in the lake. Indeed for a lake that has only existed for three decades, and which is only fed through a narrow sluice pipe, there is a surprising diversity of fish fauna. There were bream, roach, tench, pike and some extremely large carp as well as smaller species like sticklebacks. There are even gudgeon - a species usually only found in well oxygenated, relatively fast running water - and the odd goldfish which have appeared from God knows where.

Possibly the most exciting piece of news was that Pat was not the only witness to the monstrous fish. Various volunteers from the wildfowl reserve had reported seeing something large in the lake over the years, and most recently the son of the Reserve Manager - Chris Tomlinson - had actually seen something that may have just been a pair of extremely large carp spawning, but may also very well have been the elusive monster only a few weeks before our arrival on the scene.

Pat paged Chris, who came to Pat's office to meet us. A tall thin man, with laughing eyes and a baseball cap, Chris had been working at the reserve since its inception three decades ago. He, too, was interested in our quest, and described to us how the birds on the Mere had been behaving particularly strangely during the latter part of the previous winter. He told us how, each winter, the Mere (which is artificially filled so it takes up a greater area), is home to anything up to 25,000 Pink-footed Geese and 1,000 Whooper and Bewick's Swans and a host of smaller birds.

Apparently, when there is a predator in the area, such as a fox or dog, or even a human being, the birds would often take refuge in the water where they knew instinctively that they were safe from harm. However during the second half of the winter, for the first time, both men had seen the birds leave the surface of the water for the safety of the land, just as if they were frightened of something large and dangerous lurking in the murky waters beneath them. Neither of the men had ever seen anything like this before and could only conclude that there was some large, aquatic predator in the lake.

As an aside, I asked Pat whether he had heard of the links between Martin Mere and the legends of Sir Lancelot du Lac. He grinned and said that he had, and that somewhere there was a Bill Tidy cartoon showing the lady of the lake thrusting her hand girdled in white samite and clutching the sword Excalibur through the surface of the water and skewering a goose as she did it! We all laughed, and I forbore from mentioning that the lady of the lake was actually a water spirit said to inhabit Dozmary Pool in Cornwall. After all, why spoil a good funny story with the minutae of English folklore?

However, Pat's reposte stunned us. We had both been loth to mention the more peculiar aspects of the case to Pat and Chris. Although my researches over the past ten years have been enough to convince me that Zooform phenomena may on some occasions have some sort of objective reality, and that in my opinion it would be foolhardy to reject the possibility that the monster of the Mere was Zooform in nature when it was occurring in a place which had such a long and rich history of high strangeness, when Pat himself intimated that he had been thinking along the very same lines. *"Did you know that Martin Mere even had its own mermaid?"* he asked with a grin.

This was news to us, and indeed Pat didn't know much about the story, but he promised to send us relevant photocopies from a book about Lancashire folklore as soon as he could digit out. Sure enough he was true to his word, and only a few days after our return to Exeter we received a large brown envelope containing the story which has been reprinted as Appendix 1 of this present volume. Pat went on to tell us about some of the other strange stories of the Mere that, despite our diligent research we hadn't managed to unearth.

A ghostly coach was said to drive over the paths through the marshes and there were a number of other strange phantoms that had been reported over the years. There are several accounts of a malevolent water spirit or water-witch said to haunt the deeper pools of this area. As recently as the mid 20th Century There are stories of children being warned not to approach these pools because if they were to see the water-witch then they would never return home. Even discounting such

115

well known phantoms as the Grey Lady of Rufford Hall, there is no doubt that this part of western Lancashire has a rich legacy of fortean folklore. Although both Richard and I were convinced that our quarry was nothing more malign than a giant fish, it was comforting to us, to know that even if they did not believe that the monster was anything other than flesh and blood in nature, our new friends were interested enough in fortean matters to have noted some of the area's stranger folk heritage.

Now it came to the crunch. We outlined our plan of action to Pat and Chris.

1. Using an echo sounder mounted in a two man rubber dinghy, produce a contour map of the floor of the lake. Bernie had told us that there were two or three deep trenches some three metres in depth, and that the rest of the lake was fairly shallow. Chris told us that these trenches were in fact where the 1972 lake had been excavated with a JCB. Somewhere, he told us, he had some photographs of the excavations taking place. He would do his best to dig them out for us.

2. Divide the lake into three metre transects using ropes.

3. Using a sonar fish finder, search each transect in turn for sonar traces of the fish.

4. If we received such sonar traces to attempt to bait the fish to the surface using a hessian sack full of fish heads and offal. If we were successful to photograph the fish in order to produce a definitive identification. We assured them that we had no intention of harming the creature - if by some means it was possible to pull it ashore in order to measure and weigh it, that would be an added bonus, but the idea was always to release the fish back into the Mere. Pat nodded his agreement. He was a great believer in natural predation, he told us. If there is a giant wels in the Mere and it eats a few swans now and again then so be it. I, for one was so impressed by his Zen attitude to the whole affair that I would have hugged him if it wouldn't have been an appallingly *infra dig* thing to do.

So, we asked. Would they let us carry this programme out?

There was silence for what seemed like hours but was actually only a few seconds.

"Yeah, why not?" said Pat. *"It sounds like fun".* There were, of course, a few technical requirements vis a vis my public liability insurance, and the health and safety officers of the WWT, but in principle at least, we could go ahead. We could quite probably even camp on site, which was something that we hadn't even dared to ask until now. We decided that a projected time for our return would be at the end of July. Pat said that he would prefer if we did our work actually *on* the lake after the public had left. However we could spend the daytimes exploring the waterways that criss-crossed the marshes, and scrutinising the surface of the water for any sign of the mysterious creature that we were all convinced was lurking down there somewhere.

Truly, now the game was afoot, and to all intents and purposes the first part of our investigation was over.

Pat procured a pot of tea from somewhere, and we sat drinking tea and talking about catfish for a while before we made our way across the Pennines towards Yorkshire and Pat went back to have lunch with his family. We discussed the possibility that the wels may actually be a genuinely cryptozoological animal (as opposed to an out-of-place animal) after all. As Sir Christopher Lever wrote in 1975:

"John Fleming (1785-1857) includes Silurus glanis in his British Animals (1828) on the slender evidence of having discovered 'Silurus Sive Glanis' in the Scotia Jilustrata (1684) of Sir Robert Sibbald (1641-I 722) - who refers to Atlas Novus pt. 5; p.148- although Willughby, who also mentions Silurus glanis, does not list it as being found in the British Isles. It seems possible that Sibbald was describing the superficially similar burbot (Lota iota), although a number of old authorities used the term Silurus when referring to the sturgeon (Acipenser stuno)."

Lever describes more evidence for the wels being an indigenous species to Ireland:

"In his Natural History of Ireland (1849-56) Thompson, whose story is repeated by Yarrell and Day, describes the capture in Ireland of an unusual species of fish. According to Day, an unique example of a fish which some have considered may be the Silurus glanis is stated to have been captured about 1827 or 1828 from a tributary of the Shannon, near

117

its source, about three miles [5 km] above Lough Allen. A fisherman [William Blair of Florence Court] asserted that a fish at least 2 feet [76 cm] long, and 8 or 9 lbs. [3-4 kg] weight, was seen struggling in a pool in the river as a flood subsided; that it had worm-like feelers to its mouth, while its appearance was so hideous that those who first saw it were afraid of touching it. The mouth of the figure of Silurus glanis in Yarrell's British Fishes [shown to Blair by Lord Enniskillen] was not considered large enough for that of the Irish specimen, but it must be observed that inquiries were only instituted in 1840. The captured fish was not eaten but adorned a bush for two or three years until the skeleton fell to pieces, and with it all evidence to connect Silurus glanis with Ireland."

Cryptozoology is full of such unsolved riddles, and unfinished investigations. Although none of us said it out loud, all four of us made a silent vow to the rich and diverse pantheon of the Gods of Cryptoinvestigation that this quest would not be one of those sad events that started out strongly only to peter out due to lack of funding or enthusiasm. Even if we don't catch the beat on film when we return in July, we all agreed, we would amass enough evidence to be able to make a pretty firm shot at identifying the mysterious monster of the Mere.

That being said we shook hands, exchanged telephone numbers and then went our separate ways.

As soon as I got back into the car park I telephoned Graham back at the CFZ to tell him the good news. He was his usual laconic self but made enthusiastic noises. I nipped back into the gift shop to buy a present for my daughter Lisa, and then we got back into the jag and made our way East.

We drove back towards Preston in high spirits. We had achieved everything that we had intended to and more from the first half of the investigation. Now we had some serious planning to do. First however, we had to drive to Bradford to see a witch.

Our mood was in direct contrast to how it had been earlier in the day. Before our meeting with Pat it was as if everything was going wrong, but now we were in a ridiculously ebullient mood. We drove along the direct route across the Pennines. As we got nearer to Yorkshire, Richard got more and more excited. As has been recounted earlier Richard originally

118

hails from a particularly uninteresting segment of the midlands. Nuneaton seemed pleasant enough the only time that I ever went there, but for some reason Richard has always hated it. When he went to university in Leeds he fell totally in love with Yorkshire and adopted it as his new homeland. Like a Jamaican Rasta dreaming of Ethiopia he became more enthusiastic about his adopted homeland than any true Yorkshireman alive. He started to pepper his conversation with northern idioms, and after three years there he had transformed Yorkshire, in his own head at least, to being a mythical land of milk and honey where the sun never set.

When he moved to Exeter to live with me and Graham at the CFZ he brought his pro-northern bias with him. Apparently, not only is Yorkshire God's own country, but it is the only place in the entire universe worth living. After a few months in Exeter, pro-northern bias became vehement anti-southern rants. The south is not only weak and glass jawed, but is completely inhabited by those people whom Richard considers undesirables: such as homosexuals, the French, people who like Star Trek, and even the domestic cats. Everyone in Yorkshire, he proclaims regularly is rock-hard, heterosexual, keeps dogs and likes Dr Who. It is a country where it never rains, nothing is over-priced and everyone likes Goth music (like *The Cure* and *Siouxsie and the Banshees*).

All the way through our sojourn in Lancashire, Richard had been muttering darkly about the Wars of the Roses, and looking over his shoulder shiftily as if he were afraid that bands of good Yorkshire folk would stone him to death for collaborating with their ancient enemy. As we drove closer and closer to Yorkshire he became happier, and as we crossed over the border he let out a crow of delight. *"It's reet gradely to be home"*, he gushed, adopting the most ridiculous "Trouble up t'mill" voice, and exhorted me in broad Yorkshire dialect to look around at how superior the countryside on this side of the border was to that of the county that we had just left. Personally I couldn't see any difference. It was pretty enough, but to my mind not a patch on Dartmoor, but here I was the stranger in the strange land and so I wisely kept my own council as Richard ranted on about white roses.

I am very fond of Richard and am used to taking his wilder flights of fancy totally *cum grano salis*, and so I let most of his cod-northern rhetoric wash over me as we drove along through little towns, all busily readying themselves for the Jubilee celebrations.

119

For much of the time we followed the Liverpool-Leeds Canal - the very same stretch of water that we had picnicked beside a few hours earlier on the outskirts of Burscough. If our wildest hopes were to be true and the monster of the Mere, was in fact a European Catfish that had originally been released by Frank Buckland, there is every likelihood that it had originally been introduced into the Southport stretch of this very same canal. This was an exciting thought - maybe there were wels living in the entire canal. This was possibly time for a bit more historical research.

I never need much prompting to investigate the history of any canal. The great canal building craze of the 18th Century has fascinated me ever since, as a child in Hong Kong, my teacher read the class *"Friday's Tunnel"* by Sir John Verney. As readers of my inky fingered scribblings will know, Sir John Verney is one of my major literary (and socio-political) inspirations. I was lucky enough to have corresponded with him on and off for a few years before his death in the early 1990s. *"Friday's Tunnel"* was my first introduction to his work and was also my first introduction to the eccentric world of the aristocratic canal builders who squandered fortunes on ambitious waterworks which were often never finished.

Canal development in the UK was intimately bound up with the Industrial Revolution, although the earliest locked canal was built in the 16th century very near to where I live in Exeter. The first canal independent of a river was the Bridgwater Canal, started in 1759, the success of which led to a boom in canal building, which has often been dubbed The Age of Canals. Thus England and Wales had 2250 km of navigable waterways (mostly river) in 1760; by 1850 the total was 6475 km, the increase being largely due to canals. Inland water transport generally declined in the face of railway competition, and the railway companies, for various reasons acquired a quarter of the system.

The idea of linking Leeds and Liverpool by water had its beginnings in the middle of the eighteenth century and was primarily driven by the well established woollen manufacturing industry in Yorkshire. In 1770 an Act of Parliament authorised the construction of an artificial waterway between Leeds and Liverpool via Skipton, Gargrave, Colne, Whalley, Walton-le-Dale and Parbold. It was also agreed that a branch line from Parbold to Wigan would be constructed.

By 1773, under the direction of engineer John Longbottom, the lock-free section from Bingley to Skipton was completed; four years later Leeds was connected through to Gargrave but further work on the main line ceased at this point because of a lack of capital. In 1780 Liverpool to Wigan was completely canalised; prior to this time the Douglas Navigation had been used between Gathurst and Wigan .

In 1790 a new Act of Parliament was passed to raise additional funds and work recommenced on the trans-Pennine section westward from Gargrave with Robert Whitworth as engineer. Once again a shortage of money, this time caused by the Napoleonic Wars, resulted in further delays. During this period East Lancashire was developing rapidly as an industrial area and in 1794 it was agreed that the proposed line should be revised to included the towns of Burnley and Blackburn; a route originally suggested by the Lancashire businessmen. A major milestone was reached two years later with the completion of Foulridge Tunnel

In 1816, after 46 years, the longest single canal in the country was completed and became one of the most prosperous waterways. In actuality, like so many of its peers, the authorised route of the Leeds & Liverpool Canal was never completed. Rather than continue the line to Liverpool, as authorised by the Act of Parliament, the company worked out an arrangement with the Lancashire Canal Company whereby they could use the southern section of their canal between Johnson's Hillock and Wigan thereby completing the link through to Liverpool. Control of this 10 mile stretch of canal was assumed by the L & L Co. in 1864.

The branch linking Bradford to the main line at Shipley was completed in 1774 and remained open until 1922. The Rufford Branch, which connects the main line to the tidal River Douglas was completed in 1805 and remains open today. The opening of the Leigh Branch in 1820 connected the L & L main line to the Bridgwater Canal and the narrow canals of central and southern England.

Initially, limestone from the Dales area had been envisaged as the primary cargo but coal quickly superseded it. However the canal company made just as much money from carrying merchandise such as wool, grain, beer and cement. The canal was efficient and survived the onslaught of the railways in the 1880s and it was not until the lorry came on the scene after WW I that trade began to decline. The last regular commercial traffic on the main line ended in 1964 but coal continued to be carried between the Plank Lane Colliery and Wigan Power Station up until 1972.

It seemed very likely to us, that if wels had indeed been introduced into the canal at the Southport end, they might well be found in other stretches of the canal. It might be thought that we were being hopelessly optimistic about the idea that the monster of the Mere might be one of Backhand's original fish, but there are unsubstantiated claims that these aquatic monsters can live for upwards of three hundred years. One such claim is made of a specimen which died in a St Petersburg lake in the mid 1960s, after having apparently been put there during the reign of Peter the Great (1672-1725).

While I waffled on happily about canals, and Richard ranted grandiose nonsense about how Yorkshire was not only the centre of the universe, but how every single important event in history, from the renaissance to the relief of Mafeking, and from the invention of the wheel to the discovery of penicillin had been masterminded by Leeds Borough Council, we drove on happily and soon found ourselves entering Bradford. It was then that we were faced with our next problem. We had absolutely no idea whatsoever where we were going.

We were supposed to be spending a couple of days with our friends Joyce and Steve who had just bought a house in Bradford. However, we didn't have their address and neither of us knew our way around what is, with no offence meant, hardly the easiest city to negotiate ones way around in the western world. With no little difficulty we drove around a particularly intransigent one way system and found ourselves a car park. There we parked up and I telephoned Joyce and Steve on my mobile. Steve answered.

I am terribly fond of him, but he hails from the North East and has a very strong Sunderland accent. In awkward situations - say in a crowded pub, or whilst trying to talk to him down a mobile telephone whilst parked in a car park so noisy that it seemed as if *Einsterzende Neubaten* were playing a soundcheck on the building site opposite - then he is so difficult to understand that effective communication is all but impossible. After about five minutes of fighting a losing battle against the rising noise levels we finally managed to communicate our whereabouts to Steve, who then informed me in worried tones that he had no idea where the "Barton Square Car Park " was.
It turns out that Bradford is such a large and fragmented conurbation that

many people living on one side of the town have absolutely no idea of the geography of the other half. It seemed typical of our luck that we had arrived on totally the opposite side of the city to the area where Joyce and Steve had bought their new home. Steve muttered something indecipherable into his curly beard, and wandered off to find a street map of Bradford. A few minutes later he phoned us back. He had found where we were and he rattled off a string of incomprehensible instructions which I completely failed to understand. After several more attempts at explaining how to find them he gave up and told us that he would come and get us. However, Jubilee Bank Holiday or no Jubilee Bank Holiday, it was still rush hour and he instructed us not to expect him for at least half an hour.

He rang off and we sat back to wait for him. In the meantime, as I had a spare few minutes I telephoned my father.

My Dad and I love each other dearly, but over the last four decades we have not had an easy relationship. For many years one of his particular hates has been talking on the telephone. Our telephone conversations over the years have often been brief, terse and to the point. However, since my mother's death I have got into the practise of telephoning him daily at about five thirty in the evening. Somehow, and I'm really not sure why, our relationship has improved greatly in the last few months, and for the first time in years we have taken a genuine interest in each other's activities.

Every evening during our Lancashire adventure I had telephoned him as if I had been sitting at home, and I had kept him up to speed on our on-going monster hunting quest. Tonight he sounded older and frailer than I think I have ever heard him sound before. We exchanged pleasantries and he asked me whether I had been planning to attend any Jubilee celebrations. As my only plan was to get pissed whilst watching Brian Wilson of the *Beach Boys* on the Buckingham Palace Jubilee concert, I skirted around the subject, but I think I managed to cheer him up somewhat before promising to ring him back tomorrow and ringing off.

As I said to Richard, I only had the tiniest conception of what he must have been going through. I was in agony when my wife of twelve years left me. Neither Richard or I could conceive of the mental anguish that a man would go through when he was bereaved after fifty five years of marriage. Sobered by this concept we sat in silence, alone in our own thoughts, until Steve Cummins (known to all and sundry as "Curly Steve") drove into the car park and waved at us with a grin.

Nicknames are funny things. I know some people whose nicknames are so obscure in their origin that one is forced into spending an inordinate amount of time in trying to figure it out. My friend John Allegri, for example. No-one seems to know why he is known to everyone as "John-John". The bloke I was at school with who was nicknamed "Loony Fish" is another example. He was most definitely not a fish, and was one of the most depressingly sane people that I have ever encountered. Other nicknames are far easier to unravel. I am six foot seven and twenty four stone and called Jonathan. Therefore the reasons for my being known as "Big Jon" around some of the less salubrious drinking haunts of Exeter are easy to guess. John Fuller - the latest addition to the CFZ team - is black. Therefore in a community where everyone else is white and most of the blokes are called John, he became known as "Black John". It makes perfect sense. As does "Curly Steve's" appellation.

Steve is a good natured bloke in his early to mid twenties who peers happily at the world, like some woodland creature emerging from hibernation out of a veritable forest of curly hair which falls untidily over his collar, and a beard which looks for all the world like one of those odd Victorian daguerreotypes of someone wearing a beard of living bees. He is one of the nicest and most good natured people that I know, and I always enjoy being in his company.

He led us through the tortuous and twisted streets of suburban Bradford to the quiet back street where he and Joyce now lived. This was the first time that I had been to Bradford since an ill fated visit with *Steve Harley and Cockney Rebel* about ten years before when my ex wife and I had ended up sleeping rough after I took some bad acid and spent much of the concert under the erroneous belief that giant fluorescent cartoon lizards were threatening to disembowel me. As I parked the car, disembarked and staggered across (limping heavily on my walking stick) to greet Joyce, I hoped that this sojourn in Bradford's fair city would be a little less eventful.

Joyce is one of my closest friends. She was Richard's girlfriend for a while whilst he was at university,, and although the relationship fizzled out when he moved to Exeter, they remained close. From the first day that we spoke on the telephone Joyce and I have been soulmates. We can talk for hours on a wide range of esoteric subjects.

She is a few years older than me, and like me suffers from an impressive battery of health problems. She is a mother, a grandmother, an artist, author and witch. She is one of the most generous people that I have ever met and I always enjoy time spent in her company. she came forward to greet us and we hugged each other happily. Visiting Joyce and Steve is just like coming home. Poor Joyce wasn't very well. She was in the process of recovering from a particularly virulent tummy bug that was sweeping across the North of England at the time. Hotels, offices and even hospitals had been closed because of it, and quite a few planned Golden Jubilee parties had been cancelled as a result of the disease. Worst affected were children, and I learned later that there was hardly a primary school in the North of England that hadn't been affected. It was from this latter source that Joyce had contracted the disease. Her grandson Aran had come home with it, and whilst visiting his grandmother had infected her. However, Joyce was, by the time of our arrival, well on her way to recovery, and seemed extremely happy to see us.

She was accompanied by Gary, a gay friend of hers who essentially disproves Richard's claim that the entire population of the North of England is resolutely heterosexual. I said as much to Richard a few weeks later. He launched into a complicated explanation about how people from the south sneak north to try and spoil it for the rest of them. He was very unconvincing but he tried his best.

For the first few years that I knew Richard I thought that he actually *meant* these things that he says. It took me quite a long time to realise that although, for some reason, he has adopted the mantle of professional northerner, his oft stated vehemence are just part of life on Planet Richard. He just enjoys talking nonsense and being seen as a sullen, romantic, politically-incorrect outsider. It is much the same as my adoption of the Englishman abroad, with his devotion to God, Queen and Empire and insistence on Earl Grey tea and dressing for dinner in the midst of the tropics which I adopt every time I am in America. I do it because it amuses me, and I am the first to admit that I exaggerate some parts of my character, purely for effect!

Few people seem to be offended by Richard - even when he is being his most wildly offensive, and he usually seems to get away with it. I'm not sure how or why, but it seems a successful strategy for us both so long may it continue.

125

We sat in Joyce and Steve's back room, drinking tea and nibbling on an enormous pile of sandwiches provided by Gary. Our expedition was over, for the time being. We could now concentrate on having a little bit of a holiday for a couple of days before the long journey home. That evening we went out *en masse* to a little village boozer about six miles away where we met up with my old friend Phil Mantle - a UFOlogist and author of some repute. We told him about our adventures and he made us promise to let him have an account of our journeys for his new on-line magazine.

We spent the evening getting pleasantly plastered, chatting happily about things esoteric, and watching the best bits of Her Majesty's Golden Jubilee pop concert live on TV. Watching (amongst others) Brian Wilson, Steve Winwood, and a surprisingly elderly looking (and sounding) Paul McCartney singing and playing some of my favourite music while drinking happily with some of my closest friends was an idyllic way of spending the evening. When the pub closed, we bade farewell to Phil, and went back to *chez Joyce* where we drank copious amounts of alcohol and, in the wee small hours, staggered off to our respective beds for a well earned sleep.

Truly it was good to be alive

INTERMISSION

"Everything you can think of is true,
And fishes make wishes on you,
We're fighting our way up dealmland's spine,
With red flamingo's and expensive wine"

Tom Waits and Kathleen Brennan

LETTER

Mike Goldstein
GARMIN (Europe) Ltd
Unit 5, The Quadrangle
Abbey Park Industrial Estate,
Romsey SO51 9AQ

2002-06-26

Dear Mr Goldstein

In May we spoke briefly regarding Garmin Echo Sounders for fish finding. We are a non profit-making organisation, which since 1992, has carried out research all over the world into accounts of animals presently unknown to science. We are in the process of putting together the second phase of our investigation into a giant fish – probably a wels catfish – which has been attacking waterfowl (including birds as big as swans) in the Martin Mere wildfowl reserve in West Lancashire.

In May we paid a visit to the lake and spoke to various eyewitnesses, and are convinced enough of the veracity of their testimony to continue with phase two. This will initially involve using an echo sounder to explore the lake itself. We have been given permission to carry out this investigation by the officials of the Wildfowl and Wetlands Trust who administer the reserve. We then plan to bait the creature to the surface, positively identify it, photograph it and then release it again. (Complete details of the project are, of course, available should you wish to read them).

Our findings will be published in our quarterly journal, *Animals & Men*, and will then be presented to the scientific community as a whole. (Full details of our planned investigation are available should you wish to see them). The whole affair is already attracting media interest and is likely to engender a great deal of newspaper and television coverage.

We were wondering whether you would be able to either lend us an echo-sounder for a week or so, or either donate to us, or let us have an ex-display model, (for example), at a reduced cost considering the nature of our organisation and our project. We would, of course, be happy to give as much publicity as we can to your company and products, and to feature your logo and/or other company information on both our website and our printed account of our travels.

Thank you, in advance, for any help that you may be able to give us,
Yours sincerely,

(Jonathan Downes, Director, CFZ)

Tuesday 4th June 2002/
Wednesday 5th June 2002

Now, if everything had gone according to plan, we should have spent the Bank Holiday Tuesday, talking to representatives from the local angling clubs, trying to find out whether they had any records of wels catfish being caught in the easterly sections of the Leeds-Liverpool canal.

The whole day had been planned out. We were going to do that and then we were going to visit Brimmen Rocks - a peculiar valley about fifty miles north of Bradford where Richard's and my favourite children's TV series *Roger and the Rottentrolls* had been filmed. In the evening we were all planning to go out to a Pagan Moot in a nearby village, although I had a long standing promise to meet a friend of mine for dinner.

However, none of it was to be. Somehow, during her welcoming hug the previous day, Joyce had infected me with the tummy bug she had caught from Aran, and so I spent most of the day semi-delirious in bed, whilst Richard was given a guided tour of the local sights by Joyce.

The next day, though pale and shaken, I was well enough to drive, and so we made our way down south, eventually arriving at the CFZ headquarters in suburban Exwick at about 9.00 that night. Within a few minutes of our arrival Richard felt ill and he spent the next twenty four hours with the same bug that I had suffered from. However, nasty as it was at the time, it was soon over and we were able to start serious planning for the next leg of the operation.........

The whole thing was my mother's fault really. When I was a very small child she introduced me to the books of the late Gerald Durrell, and from a very young age he became my greatest inspiration. After a childhood spent as an avid amateur naturalist, and a short apprenticeship spent at Whipsnade Zoo, he embarked on a series of animal collecting expeditions in Africa and South America during the late 1940s and 1950s. He started his own zoo on the island of Jersey in the year I was born and a few years after turned the whole thing over to the Jersey Wildlife Preservation Trust – a non profit making organisation which has achieved marvellous things over the last four decades and which has been responsible for saving a number of species from otherwise certain extinction.

I suppose that I always wanted to do the same thing. However, I was well into my forties before I had the chance. At the time of writing I have been a professional fortean zoologist for over a decade. Like everything else in my life it happened almost by chance. After a serendipitous chat in 1994 with a representative of one of the better known Sunday newspapers about the "Beast of Exmoor" and the subject of Cryptozoology - the study of unknown animals - a subject that had been my hobby for many years, I was quite surprised to read that I was "one of Britain's foremost cryptozoologists".

In a 2001 interview with the Yorkshire UFO Society I was asked how I first became involved in the subject:

"I've been interested in this stuff all my life. My interest probably started when I was a little boy. I was brought up in Hong Kong surrounded by all sorts of weird and wonderful animals anyway. Every week my mother used to get me library books and one week when I was about seven years old she got me a book called "Myth or Monster" which introduced me to the wonderful concept that there were real monsters like the yeti, the loch Ness monster and sea serpents in the world around us. I was immediately hooked and I've been interested in Cryptozoology ever since."

Cryptozoology is the study of hidden or unknown animals, and such creatures, belonging to a species wholly or partly unknown to science, are, as explained earlier, usually collectively referred to as 'cryptids'. Pseudo-cryptids and zooform phenomena are, really, nothing to do with cryptozoology in its purest sense.

As I became more deeply involved in the study of Zooform phenomena I

130

began to realise that you could no longer study these 'creatures' in isolation. In many cases, Zooform phenomena are inextricably linked with a wide range of other paranormal and fortean phenomena, most especially crop circles and UFOs.

It was because of this curious dichotomy of definition that my ex-wife and I finally decided to found the Centre for Fortean Zoology in April 1992. We were staying with friends in Derby and discussing the problem of my being a cryptozoologist who didn't necessarily want to study Cryptozoology per se when my friend Dave said, *"Well, I guess you need to start a new science then."* :

I had to agree with him. I was already a devotee of the work and philosophy of Charles Fort, the American researcher who had given his name (unwittingly and one suspects unwillingly) to the study of anomalous phenomena and so the name Fortean Zoology really found itself.

No, that's not entirely true.

I originally wanted to name my nascent scientific discipline Anarchozoology. Taking the political philosophy of anarchism, rather than the popular interpretation of the term, my concept of Anarchozoology was to be a discipline where one made up one's own rules, and then stuck to them, rather than being bound by the rigid dogma of the preconceptions of someone else's scientific world view. It seemed to me then that the Omniverse was such a strange and bewildering place that it was only by fusing the two apparently disparate concepts of 'discipline' and 'anarchy' that one could make any sense of it all. It still seems like that to me today.

I had been working as a freelance writer and researcher for some years, and it seemed to me that the time was ripe to formalise my research, and to form an institution through which I could share my research with others working in similar disciplines.

However, this was not the full reason. Because of the very nature of mystery animal research, and especially when one is considering the subject of Zooform phenomena, which in itself is a strange and disturbing discipline, there is a tendency for some sectors of society to treat you like a bloody lunatic because you date to show an interest in things that are away from the norm. Perhaps the fact that at the time I was a drug addled

131

hippie didn't really help, but I felt that I needed all the spurious credibility that I could get if we were going to have any chance of being taken seriously researching into such diverse subjects as frogfalls, Cornish Big Cats, animal mutilations, and the Owlman of Mawnan (all of which we looked into before the CFZ was a year old)

Although when it started the CFZ consisted of Alison, me and my mate Dave from Derby, after a couple of years it seemed a logical move to start expanding our horizons. Serendipitously, at the time I had recently made friends with a lady called Jan Williams from the North of England who, together with Trevor "Beast of Exmoor" Beer was running a short lived organisation called S.C.A.N - (The Society for Cryptozoology and the Anomalies of Nature). I became a regular contributor to their magazine, and when after four or five issues, Jan and Trevor went their separate ways I asked Jan, diffidently, whether she fancied helping me start up a Cryptozoological magazine. I wanted to call it *Animals & Men* after a line in a song by *Adam and the Ants* from their unjustly ignored debut album *Dirk Wears White Sox*. Remember at the time I was still a bloody hippy and my writings and methodology were horribly imbued with references to anarchic politics and obscure rock records. Much to my eternal gratitude Jan agreed and the magazine (and thus the CFZ proper) was born.

The rest is history!

Alison and I were practically bankrupt at the time that the first issue of *Animals & Men* was published in the early summer of 1994. We hitchhiked to London and gatecrashed the first ever *Fortean Times* Unconvention with our complete stock of two hundred of these magazines. Much to my surprise I sold the lot and sat at the bar feeling rather please with myself

At the bar was a middle aged Irishman with a bodrhan, wild staring eyes and attitude. He was Tony "Doc" Shiels, and he and I soon got talking and drinking. Within an hour we were firm friends. A journalist from The Guardian came up to me and asked me and my companion for an interview. We were both pissed and slightly belligerent. He asked who we were. "Doc" threatened to curse him, and I replied in my most pompous ex public schoolboy voice. *"Dear Boy, I'm Britain's best known Cryptozoologist"......* Of course I wasn't any such thing, but they printed it anyway, and if its in the papers it must be the truth 'cos seven years later I'm still here and we've just published issue twenty four of Animals & Men. Funny old world innit?

By the end of the weekend we had well over a hundred subscribers and our future was secure. We continued publishing magazines at roughly three monthly intervals but it wasn't until issue three that we had our first proper break.

Also at the 1994 Unconvention I met Dr Karl P.N.Shuker, someone with a REAL claim to be Britain's greatest cryptozoologist at the time, and someone else who has, over the years, become a good and dear friend. He was lecturing and during his lecture he told the fascinating tale of some alleged monster footage that had been taken at Lake Dakataua on the Island of New Britain off the coast of New Guinea. This story galvanised the entire British cryptozoological establishment and we dedicated many pages of the next few issues to the story until just after issue 4 we managed to get hold of a copy.

We were the first people in the UK to get a copy, and indeed (sorry to those awfully nice fellows at Tokyo Broadcasting Company TBS) but as far as we know all the video copies currently circulating in the UK originally came from us. Unfortunately for cryptozoology as a whole the pictures turned out to be of salt water crocodile(s) but fortunately for us, not only did it take some time for this fact to be discovered but it was one of our researchers, Darren Naish, who discovered the fact and first published it in our 1997 Yearbook.

In 1995 we decided to start a second publication, the annual Yearbook. This 200 plus page collection of longer research papers has been published annually ever since.

Unfortunately in 1996 Alison and I separated and were divorced after long, messy and painful legal proceedings a year or so later. My old friend and Colleague Graham Inglis whose main qualifications for the job were that he was:

a. An old mate
b. Had some knowledge of administration and computerised office pro cedures
c. Had worked with me on a number of projects over the years
d. Could party like it was 1999 when it was only 1996

came on board to fill Alison's roles as administrator and general partner in crime.

Even now, six years after the split I look back at those months with horror and I can truthfully say that if it had not been for the strength and fortitude of Graham and my many other friends in the CFZ around the UK I don't think I would be sitting here now writing this story. I would probably be dead. And I mean that.

In the mid 1990s the Caribbean island of Puerto Rico was plagued by an outbreak of animal mutilations. These were blamed on El Chupacabra (Spanish for The Goat Sucker - also a derogatory term for prostitutes of the lowest order).

The reports described attacks on a wide range of domestic livestock and there were even disturbing reports of attacks on human beings. Researcher Conrad Goeringer wrote in 1997 that:

"Believers in the chupacabras say that the beast is a hybrid creature, in appearance something which resembles a cross between a giant dog and a lizard. It is said to walk upright on two feet, is capable of flight, and sinks its fangs into victims and kills them by drinking their blood. News reports of chupacabras sightings come from mostly rural areas; and while the mysterious creature seems to prefer farm animals like sheep, goats, and chickens, it has been alleged to attack humans."

A researcher at the Centre for Fortean Zoology, who shall remain nameless described the animal (most famously depicted by Puerto Rican researcher Jorge Martin) as a cross between a kangaroo and Sonic the Hedgehog (a computer game character) on acid! By 1996 the attacks had spread to Mexico, Guatemala and even the mainland United States.

The reports continued and in September 1997 we were approached by AVP Films, an independent company to take an expedition to Puerto Rico, Mexico and Miami in the hope of tracking down some witnesses and discovering the truth about the creature.

The resulting adventures are told in the Channel 4 Film The Fearless Vampire Hunters and in my book Only Fools and Goatsuckers which was belatedly published by CFZ Press in 2001..

134

On our return to the UK Graham and I were only too aware that if we were to expand our activities we needed to take someone else on board. At that time we had been corresponding on and off with a geezer called Richard Freeman for several years. He had co-authored an article on Dog Headed Men with me for the third Volume of *Fortean Studies* and we had collaborated on a number of short TV shows. He, too, was a party animal with a stupid sense of humour, but unlike Graham or me he had proper zoological qualifications. He had followed up a City and Guilds in Animals Management and he had studied zoology at Leeds University.

We asked him whether he wanted to join us at the CFZ. He did and he is still here four years later.

All the pieces were now in place. By the end of 1998 we were all living in slightly bohemian squalor at my house in Exeter. The CFZ was now in place and all were present and correct.

For a variety of reasons, which I don't feel the need to go through here, my personal finances had improved by the early months of 2002 to such an extent that I felt able to take the first tentative steps towards making the CFZ an official charity. It was becoming increasingly clear to me that if we were to carry out all the work that we wanted to do, we would need to get hold of some proper funding. Over the years the CFZ had been funded purely by what the three of us could earn through writing and television work, by the sales of our books and magazines, by the revenue generated from our annual conference (*The Wyrd Weekend*) and from my Disability Benefit.

This was no longer enough to allow us to expand in the directions that we wanted to, and so, as I was by now earning enough money to be relatively financially independent, we took the plunge, and with the assistance of one of our subscribers (the long suffering Martin Jenkins), we drew up a constitution and became registered as a non profit making trust.

One of our first acts was to take on a new addition to the team. In his election manifesto that persuaded the gullible British people to elect him to power in 1997, The Right Hon. Tony Blair MP had promised that all people who were long term unemployed would be offered work experience places with various suitable employers. I am absolutely certain that when he drafted the legislation which put this portion of the "New Deal" (even the name had been taken from an economic drive sponsored by the US Government between the wars), into practise he never thought that this would allow, a young man called John Fuller – half English and half Ethiopian and known to everyone as "Black John" – to join the CFZ at the taxpayer's expense in order (on paper at least) to learn the intricacies of filing, computer typesetting and interminable use of our photocopier. However, thankfully for the CFZ this is exactly what happened, and at the beginning of May 2002 John joined us.

We soon realised that what was originally planned as a simple three month arrangement was going to turn into a full time arrangement, as John fitted happily into the organised madness that is the CFZ at its shambolic best. On our return to Exeter after our north country anabasis, we started making plans for our return. Without even bothering to consult each other, Graham, Richard and I took it for granted that John would accompany us when we returned to the lake. It seemed that John took it for granted also, because within days of our return, it was firmly established that he would be accompanying us.

A couple of days after we came back a large A4 manilla envelope appeared on the CFZ door mat. It contained the photocopies of the "Mermaid of the Mere" story (see Appendix One) which Pat had promised us. We read them with avid interest, and added them to our ever more voluminous files on the high strangeness that seems to have surrounded the Mere for centuries.

LETTER

2002-06-27

Dear Pat,

Thank you very much for the mermaid photocopies that you sent me. They only go to prove, (as if any proof was necessary) that Martin Mere has the potential to be quite a strange place. I have collected an enormous amount of folklore about monsters and dragons from what used to be the marshes surrounding the ancient boundaries of the lake, and they will all, be in my book when it finally gets written.

I am sorry that it has taken so long for me to get back to you, but I have had a few family problems. However, with your permission we intend to come back up to Lancashire during the last week in July to conclude our search for your mysterious beastie.

Is this OK with you datewise? Also, have you had a chance to sort out the Health and Safety implications yet? As we discussed during our meeting on the Jubilee Bank Holiday, if your H&S officers want me to take out insurance cover for the two people who will actually be *going* on the lake, I shall be happy to do so.

We also discussed the possibility of us camping overnight to save on expenses. What I was wondering, is, would we possibly be allowed to bring the cars through and park by one of the hides, and spend the night in the hide (ideal for using night sights on the surface of the water – these bloody catfish are night feeders) and then pack our stuff into the car before the punters arrive in the morning.

We would spend the day searching local museums and tracing the drainage ditches from the canal, and then, as discussed, do our aquatic activities with you after the punters have gone home during the long summer evening.

If this is all OK with you, we will be extremely grateful.
Best wishes,

Jon Downes

LETTER

3 July 2002

Dear Jon,

Thank you for your letter of 27.6.2002. Apologies for not replying sooner due to my being on holiday. The last week in July will be OK, though a number of our staff are on holiday then.

As regards the **"formal'** side of things

1.   I enclose a risk assessment for the use of boats on our sites. You will be required to abide by the rules of the risk assessment
2.   It will be advisable for you to take out insurance cover for the two people on the boat. You may also wish to take out third party cover in the (albeit unlikely) circumstance of a member of the public being injured by your equipment / car etc.
3.   My Health and Safety officer will need to inspect the boat / equipment to ensure it is suitable for the job.
4.   It would give me some "comfort" if you would write back with a disclaimer saying that we, WWT, will not be held responsible for any accidents, injury or illness which may result from this exercise.
5.   Finally, I would need a written method statement which sets out clearly and in chronological order what you intend to carry out during your visit. This is something that we require of contractors working on site to satisfy the law.

Sorry to be so fussy but there is a "history" to all this which I may explain to you some time.
I see no problem in you bringing your vehicles inside the perimeter fence overnight but we can discuss the practicalities when you arrive.

Hope that covers everything.

Best wishes.

Yours sincerely,

Patrick J. Wisniewski B.Sc. M.Sc.
Centre Operations Manager

RISK ASSESSMENT DOCUMENT

WWT RISK ASSESSMENTS

DEPARTMENT
Grounds 1 Conservation

TASK
Working in water over normal wader depth and in boats or canoes

HAZARDS
Deep water (up to 3m depth) with uneven pond base. Water and mud
combination (suction) low temperature, some work is carried out in
ice conditions.

WHO IS AT RISK

Grounds staff, Volunteers, contractors

POSSIBLE EFFECTS
Drowning, Becoming trapped in mud

MEASURES TO CONTROL RISKS
A minimum of two people to work together both with life jackets and
radios, one to act as safety backup for the other. Safety line
should be available if needed. Boats should only be used in Trust
grounds, not for outside rescues etc. Spray decks should only be
used on canoes by trained personnel Life jackets must be worn in
boats/canoes. All cuts should be covered before work in water and
hands/face washed after work with anti-bacterial soap. Boat users
must be capable swimmers.

SAFETY EQUIPMENT

Life jackets. Safety rope. Radios.

IS THE RISK CONTROLLED   YES

OVERALL SCORE 7/10

NOTES

The risk is controlled in the Trust waters. Work in other unknown
water would be an on the spot risk assessment

Date 3.7.02 Signed: P Wisniewski

Everything was okay and we could do exactly what we wanted. The game was most definitely afoot.

As preparations for the second leg of the expedition continued apace, the long awaited documents containing the Deed of Entrustment finally came back from HM Stamp Office. The next day, Richard, Graham and I went out and got ourselves a CFZ building society account, and commissioned a new all singing all dancing logo from our artist Mark North.

I can't remember whose idea it was (but I think it was mine), but we decided that it would be a good idea from the PR point of view if the motley collection of middle aged men who were to make up the CFZ expeditionary force to Lancashire in July were to wear some sort of uniform. I discussed this with Tim Matthews on the telephone and he was almost incoherent with laughter as he begged us to allow him a photo opportunity of him outside Martin Mere nature reserve inspecting the serried ranks of CFZ personnel in full battledress.

"Can you imagine what O'Hara would do with a picture like that?" he giggled down the telephone.

I must admit that the amusement factor of the situation was not lost on me, but on a serious note, as we were going to be stationed at the Nature Reserve itself for several days, if the CFZ posse were wearing some kind of uniform, it would not only make for a striking image in publicity photographs, but might engender a fair amount of public interest, and hopefully some more support for our activities.

Working on the principle that all publicity is good publicity we decided on a uniform of battledress trousers and jacket and a black T Shirt emblazoned with the new CFZ logo.

Tim Matthews, we felt, and probably Gerald Durrell, would have been proud of us!

140

LETTER:

To whom it may concern,

Re: CFZ's lake contour mapping and search for fishes, 25th – 30th
July 2002

I confirm that you (ie the Wildfowl and Wetlands Centre) will not
be held responsible for any accidents, injury or illness which may
result from the above-mentioned exercise.

This document indemnifies the Wildfowl and Wetlands Trust, all its
employees and associates in perpetuity against responsibility for
any such events

Signed

Jonathan Downes
(Director CFZ)

Graham Inglis
(Deputy Director CFZ)

TRUSTEES: Jonathan Downes; Graham Inglis; Richard Freeman. Registered as a Trust with HM Stamp Office

Our next job was to gather together the specialist equipment which we would need to complete a job which was entirely different to anything that we had ever attempted before.

I am always impressed when you read tales of high adventure. The protagonists always seem to have plenty of money and large stocks of specialist equipment to hand. You can never imagine them fannying about trying to get things on the cheap from the local free papers, or negotiating earnestly with a drug dealer from Yeovil trying to get twenty quid knocked off the price of a rubber dinghy. However, unlike all the other people who seem to write books about their adventures, the CFZ is always strapped for cash and several of its most hard core members are actually on the dole at any particular time.

We are therefore forced into situations that other adventurers cannot even dream of, and which in many ways are often an adventure in themselves. Until now most of our exploits had been land based, and so we were ill equipped to take on an investigation which would perforce necessitate much of the time being spent afloat. In fact when Graham and I tried to construct an inventory of all the nautical equipment available to us we found that it consisted of a pair of wellies, an old landing net and a bag of rusty golf clubs. The rusty golf clubs do not really count, but because they were stuck inside one of the wellies, and the landing net was torn, so our actual tally of maritime equipment came down to one pair of wellies which my ex-wife had got from a jumble sale and which didn't actually fit anyone presently in the CFZ posse.

The CFZ is presently housed in a small mid terraced house which was built in 1985 as a starter home for young married couples who hadn't got enough money to buy anywhere decent. It is built up against the side of a hill which means that although there is, on paper at least, a quite sizeable expanse of garden most of it is at a 45 degree angle up a slope covered with brambles, and what is left is a dingy piece of land about twelve feet square. Because we needed more room far more than we needed a dingy piece of land twelve foot square, we have built a rickety shanty onto the back of the house. This two-roomed edifice (which I pompously refer to as the conservatory to anyone who hasn't actually *seen* it) contains a fibreglass pool in which I try to keep goldfish and do my best to deter Helios 7 (one of the three CFZ pussy cats) from voiding her bowels into, an enormous tank which most summers is home to a large and manky looking alligator snapping turtle, and an assortment of filing cabinets. It also contains a variety of strange objects which the CFZ have used in our investigations over the years.

The fact that at the moment it is chock full of semi-maritime type stuff is testament to our incredible powers of, if not getting something for nothing, getting quite a lot for very little. Every Tuesday and Wednesday a series of newspapers are published which contain adverts for a wide range of impedimenta from books to cars, and from horses to computers. For some reason the versions published in the westcountry at least are printed on canary yellow paper, and are known colloquially as the "yellow papers". It was here that we first went in search of all the equipment that we figured that we were going to need, and, not surprisingly, it was here that we found most of it.

The first two things that we needed were a fishfinder and a boat. We had written both to the manufacturers of fishfinding equipment and to the most well known vendors of the same, in a vain attempt to blag one for free, or at least cheaply. We decided therefore to concentrate our efforts first upon finding a boat. It was obvious, fairly quickly, that this was going to be somewhat of an onerous task 'cos boats (even rubber dinghies of the type that we had envisaged), were, at the beginning of the summer holiday season mostly prohibitively expensive.

We had been looking for a suitable boat for about a month and time was getting short, when one Tuesday we found what appeared to be an appropriate craft advertised in the free-ads.

"FOR SALE. 4 Man Rubber Dinghy as used by coastguard. Paddles, puncture kit, only used once. Genuine reason for sale. £50.00 ono"

I have always found the line "Genuine Reason for Sale" highly amusing. The 'genuine reason' is so often that the item in question is a piece of crap or that the bloke selling it is strapped for cash. Sometimes the 'genuine reason' is that the item is either broken or stolen and sometimes there is an even more exotic explanation for why some poor geezer is reduced to selling his worldly goods to a complete stranger through the somewhat sordid medium of a newspaper's free small ads. The advert gave a Yeovil telephone number so we rang up the bloke concerned who was apparently called 'Dazza' and arranged to go and view the boat.

The fact is that we had already made up our mind to buy it. It was the only one that we could afford and if it was half as good as the description printed in the free-ads, which had been reiterated by Dazza on the telephone then we were onto a bargain.

So we dropped what we were doing and Graham drove us to Yeovil in the old CFZ Volvo Estate. Yeovil is a spectacularly unattractive town about fifty miles from Exeter and it has always been a place that I have done my best to avoid like the plague. I played a few gigs there back in the days when I was touring incessantly with my musical ensemble, and I had grown to dislike the place intensely. We drove through the outskirts of what had once, probably been a pretty little Somerset market town, but were now internecine industrial estates covered with graffiti, and did our best to follow the instructions that Graham had scrawled on the back of a cigarette packet.

We eventually found the place where "Dazza" was supposed to live, and on first sight it appeared to be a highly unlikely location for any vendor of rubber dinghies. It was a small group of blocks of flats of a semi-institutional nature, at least part of which seemed to have been allocated to a group home fort the mentally handicapped and the rest seemed to be sheltered accommodation for old people. We parked up in a car parking area which was labelled "Physiotherapists ONLY" and looked around for some guidance. A middle aged man with Down's syndrome ambled towards us grinning happily to himself. He was holding hands with an elderly gentleman wearing distinguished looking tweeds which gave him the appearance of the late Harold Macmillan, and who sported the distinctive stigmata of a prefrontal lobotomy on his forehead. Preferring the advice of people who had the requisite number of chromosomes, and were still in possession of their frontal lobes we decided to ignore them and do what we should have done earlier. We telephoned "Dazza" on Graham's mobile phone.

The two men gazed into the back window of the estate car with rapturous delight and started to fondle each other as we explained to "Dazza" that we really had no idea where we were. He told us to look out of the car window and up, which we did, to see a very disreputable looking person in a pair of blue and white cycle shorts leaning over the balcony of a third floor flat and shouting *"Geezah"*.

This, apparently was "Dazza". He was neither geriatric or mentally handicapped as far as we could see and, then at least, we could see no reason why he would be living in a place which was obviously dedicated to serving the needs of the more vulnerable members of the Yeovil community. It was only half an hour later when his girlfriend appeared that everything became clear. She was one of the most stunningly beautiful young women that I have ever seen and she couldn't have been more than seventeen, but she was also equally obviously - how does one put it in words that are going to be acceptable to the politically correct of my readers? - several sandwiches short of a picnic.

It turned out that it was her flat. She, despite her beauty, was obviously well within the targeted client group of the place, and "Dazza" was her boyfriend. It also became very obvious very quickly that he was not only living in her flat but that he was using it was a base for a flourishing drug dealing operation. The house was spotlessly clean and furnished with exquisitely tacky bad taste. From my knowledge of things, having worked for some years as the Assistant Team Leader in a group home for the Mentally Handicapped I know only too well that whoever it is who has the job of furnishing and decorating these places, they have absolutely no taste whatsoever. The wallpaper and soft furnishings were garishly coloured and clashed appallingly. Whoever the purchasing department of the local council were, they obviously had never heard of the slightly Zen concept of providing a harmonious and peaceful environment for people with emotional and mental problems. Just staring for five minutes at the clashing colours and the ugly furniture gave me the beginnings of a headache. God only knows what effect it would have on some poor mentally handicapped girl who had to live there.

My house is too small and has too much stuff in it, but I don't think that

I have ever seen so much tacky rubbish packed into such a small place in my life! I have often wondered who it is that actually *buys* all the over-priced commemorative plates and collectible porcelain figurines that are sold through the colour supplements of the Sunday newspapers. Now I knew. Everywhere you looked there were display cabinets full of col-lectible china thimbles, portraits of Elvis painted in dayglo paint on black velvet, porcelain kittens sporting sycophantic grins and plates com-memorating the high spots of forty years of *Coronation Street*.

However, there was also a large pair of brass scales on the coffee table, and a small bundle of plastic bags containing pills and what were either oxo cubes and dried oregano or hashish and cannabis leaf. Although usu-ally if I had been faced with a situation like this I would have run a mile, beggars couldn't be choosers and we needed a boat very badly. Unfortu-nately when we saw the boat it was not anywhere near as amazing a bar-gain as we had originally been led to believe. It was about eight foot long but as far as we could see the only connection that it had with H.M Coastguard was its smart colouration of black and yellow. How it could ever fit four people was completely beyond us, but it would fit two peo-ple in it in reasonable comfort which is all we would need it to do, and the drug business was obviously doing well and "Dazza" was prepared to take forty quid for the thing and so we bought it and made a fast get-away.

It was only when we got down to the car with the deflated boat and the bundle of oars, that we were faced with our next problem. The two rear tyres of the Volvo were completely flat and there was a small gaggle of elderly women with Sturge Weber's syndrome trying to hide behind a completely inadequate poplar tree and giggling behind their hands. We had practically no call credit left on our mobiles, so we had to traipse up the flights of stairs to "Dazza's" flat again, and ask whether we could use his telephone to summon the rescue services.

To a lot of people of my age or older, short cropped hair has certain cul-tural connotations. However times have changed and these haircuts do not any longer mean that the person wearing it is an unpleasant racist thug and/or football hooligan. Quite a lot of my friends (even those who aren't unpleasant racist thugs and/or bootboys) sport such haircuts these days and most of them are very nice people. However, when the afore-said haircut is worn in conjunction with a tattoo saying "A.C.A.B" ("all coppers are bastards" for the uninitiated), and another one along his neck

147

of a dotted line and a pair of scissors with the caption "please cut here", then I don't think that even my severest critic would accuse me of bias when I say that I looked at him askance.

When this person is wearing a Tellytubbies T Shirt and is standing next to a person whom we were only too aware was selling class A drugs to an estate full of mentally handicapped adults who were deemed to be in need of care and protection by the state one is even less likely to want to have anything to do with him. However, this bloke (who turned out to be "Dazza's" brother "Tezza"), was actually incredibly helpful. The two unsavoury characters couldn't have been more helpful

They produced a foot pump and within a quarter of an hour we were back on the road. "Dazza" and "Tezza" exhorted us to come and visit them again soon (something that we were all determined hat we would never do), and we drove back to Exeter as quickly as we could. Looking in the rear view mirror as we drove out of the little close we could see the football hooligan, the drug dealer, and a gaggle of their subnormal customers waving goodbye to us. In a long career of having visited some strange and unpleasant places this was probably the strangest and we were very glad to leave.

But at least we had a boat. Graham had some structural alterations to make to it and over the next week or so he added a plywood deck, a seat, and a wooden superstructure which pointed out over the bows, upon which the transducer of the fishfinder would hang as we navigated our little craft over the surface of the lake.

The next problem was the fishfinder. We didn't have one.

Garmin International took no interest in our proposal, and they wrote me a letter which politely but clinically refused our offer. The bloke from Marconi, although he had shown a great deal of interest in the project when we spoke to hi on the telephone didn't even bother to answer my letter. The fishfinder was becoming a major obstacle to us, and there was no way that we could afford to buy a new one. The cheapest models re-tailed at about two hundred and fifty quid inclusive of VAT and that was well over our price range.

So it was back to the yellow papers.

For two weeks there was nothing and we were beginning to panic. Then, with just over a week to go until we were scheduled to return to Martin Mere it all came together. In one issue of the free-ads there was a fish-finder, a pair of waders, some lifejackets and other stuff that we needed for the expedition, but there was only one problem. In order to schedule sufficient days off for Graham and John before we embarked on our great adventure we only had two free days, and one of them was going to be taken out in doing live trials of the equipment down on the River Exe.

After a frantic night of making telephone calls to arrange appointments we set out early one Tuesday morning. Graham was driving the Jag, and Richard and Lisa were in the back seat. We had decided to kill as many birds as possible with the fewest possible number of stones and so we had scheduled in a ridiculously complicated schedule for the day ahead. It does not need the benefit of hindsight to know that this was potentially a horrible mistake. We knew that it was at the time but there was abso-lutely nothing that we could do about it.

What made the day even more complicated is that this was the day that my father went in to hospital for an operation on his knee. My father does not like people to fuss over him while he is in hospital, so I never let him know that I am fussing although I always keep in touch with the Charge Nurse on the ward he is on each day to find out how he is. Ironi-cally our first port of call was a little village just north of Barnstaple which meant that we would have to drive right past the North Devon District Hospital at Pilton, at about the same time as my father was un-dergoing his operation.

I spent many of my formative years in North Devon, and many of the places that we were due to visit that day had poignant memories for me. But this was one of the first times that I had been back to Barnstaple for nearly twenty years, and it had changed out of all recognition and soon we were terminally lost.

In the early 1980s I had been pivotally involved with the alternative/ punk/new wave scene in the town which revolved around two geezers called Lydon Garner and Colin "Bunker" Brazier. They travelled around in an old blue ambulance and owned a little record company called 'Next Wave' records which released several singles. The scene was based in a number of the smaller and seedier pubs in town, especially one called *The Royal Norfolk* which used to stand down opposite the bus station. I

149

had been looking forward to revisiting the haunts of my youth, and during the long journey north along the North Devon Link Road (NDL) I had regaled my travelling companions with long, boring stories about my younger days as a spikey haired punk rocker, and whilst Lisa listened kindly (humouring her old adopted dad in the way that daughters have done throughout the centuries) the other two paid very little attention.

When we got to Barnstaple it was a great disappointment as I could hardly recognise anything. *The Royal Norfolk* had been replaced by a garishly painted night-club, the back street pubs where I had plotted world domination with a coterie of similar spiky haired 20 year olds, were now boutiques and wine bars, and the rabbit warren of tiny streets and Victorian houses which had once been home to an entire culture of musicians, poets, fanzine editors, promoters, singers and painters had fallen foul of the planners and was now a DIY superstore and adjacent car parking facilities.

"Screw this", I thought, *"maybe you should just never go back"*, and I navigated us through the town towards the little village of Muddiford where, according to the trusty yellow papers and confirmed by the alleged vendor, there was a sonar fishfinder waiting for us for the paltry sum of a hundred and fifty quid.

This first leg of the day's activities went surprisingly smoothly. We managed to negotiate the twisting North Devon lanes that I remembered so well from the days of my youth. We found the vendor's house without any great difficulty, and knocked on his door. If the middle aged Chartered Accountant who was trying to sell a fishfinder that he had bought a few years before on a whim and had never used, was surprised to see our motley band disembarking from a dark blue Jaguar outside his front door he didn't show it. The advent of a hairy fat man with a walking stick, a plump balding Goth doing his best not to leer lasciviously at the vendor's extremely attractive and well stacked blonde daughter, and a painfully thin man with wild, staring eyes seemed to faze him not at all. Lisa wandered over to make friends with a small gaggle of Chinese geese that gobbled and gurgled at her through the fence of the small paddock on the opposite side of the road, and the rest of us went in to the vendor's house to look at the fishfinder.

Unfortunately he seemed to know next to nothing about the machine. He didn't even know whether it worked or not, but he rather grudgingly produced a car battery and some lengths of wire and looked on impatiently as Graham connected the thing up. I have to say that I didn't warm to him particularly. He reminded me too much of the bad tempered little housemaster who had made my schooldays at the minor public school only a few miles away at which I had spent a few unhappy terms in the mid 1970s. Graham, quite properly, wished to ascertain whether or not the thing actually worked, and the vendor just wanted to sell it, so as he fulminated and bristled to himself, Graham conducted what tests he could, ascertained that to the best of his knowledge it was indeed working and gave me a nod. With a thin smile I gave the little man £150 in notes from my breast pocket and then we returned to the car to resume our journey.

This is when things really started to become weird.

The first objective of our journey had been achieved so easily that I decided that we could afford to take a little detour. When I first came to England in the spring of 1971, my family lived for a while in Wiltshire before moving to Woolsery - a small village in North Devon. My father still lived there thirty one years later. My brother and I had become close friends with the three children who lived next door. David was about the same age as me, and his sisters Lorraine and Kaye were two and four years younger respectively. We remained friends over the years even after David's tragic death in 1987. When Kaye had her youngest child Greg in 1998 she asked me to be his Godfather. This was a task that I was happy to take on. I had been working, during 1999/2000 on a children's animal encyclopaedia and had managed to blag a copy from the publishers. All four thick volumes were now residing in the boot of my car after having been left on my landing for an appalling length of time. I had promised them to my godson and his brothers and had been waiting for some time for an opportunity to deliver them.

The previous occasion I had been to the village where my parents had lived for the previous three decades had been on the occasion of my mother's funeral, and I had not been in any fit state to make house calls but today was different, and as the Jaguar, in Graham's capable hands sped along the well remembered roads towards my childhood lane I leaned half around in the passenger seat and regaled Lisa with stories of my boyhood. She listened politely, but I don't think that she was any more interested in them than I had been when my father told much the same stories to me.

I tried to telephone Kaye on a number of occasions during the eighteen mile journey between Barnstaple and the village we had both grown up. There had been no answer, but we drove ahead anyway. It was only as we approached Woolsery that I realised that there was a fundamental flaw in my plan. I had absolutely no idea where she lived. She had moved house a few years before and although I had her telephone number I didn't have her address. I didn't know whether she had a mobile phone, and even if she did I didn't have the number. We parked by the village shop and I felt horribly out of place. This was the place, where, in the early 1970s, I used to go every morning during the summer holidays, armed with my butterfly net, to see what moth species had been attracted overnight by the streetlights and were roosting on the white plaster walls. In thirty years nothing much had changed, but the kids from the village school were going home with their mothers and I was acutely aware that the sight of three slightly strange looking middle aged men sitting in a Jaguar watching the school gates (I guessed that Kaye would have had to go to pick up the kids), must have raised alarm bells in the minds of some of the women who saw us.

The place may have looked the same, but the socio-political climate had changed beyond all recognition. For some reason I felt uncomfortable in my old home village that afternoon and so I was actually quite relieved when, the animal encyclopaedias still in the boot of the Jag, we continued our journey down towards the Cornish border.

Again these roads were only too familiar to me. This was the stretch of road upon which my father had taught me to drive in the late 1970s but that had been a bad time for me, and indeed one of the times when my relationship with my father - now languishing in a ward at Barnstaple hospital - was at its lowest ebb and therefore, in a stark contrast to my high spirits earlier in the day, this leg of the journey was not particularly enjoyable for me.

The next stage of the journey was just plain weird.

It took over an hour for us to make the journey from Woolsery to Wadebridge. This is not a town I know particularly well although I have visited it on occasion over the past thirty years. Once again, as we were approaching the town therefore I gave the man whom we were supposed to see a ring on the mobile. The number came back as unobtainable. Now, this next appointment was almost as important as the fishfinder had been. It had been stipulated by Pat Wisniewski that life jackets would be worn by all CFZ personnel who went in our little dinghy. Even if this stipulation had not been made, I would have insisted upon it myself, and so we were in desperate need of life jackets.

What I did not realise, however, was quite how expensive the damn things are! The best price that we had been quoted was fifty quid each, and so when we saw a pair advertised in Cornwall for thirty quid there was nothing that we could do except to travel to Wadebridge to purchase them. There was only one problem. Although we had the address of the place where we had to go, no-one had the slightest idea of where the hell it was.

So we did what any self respecting explorers try to do under these circumstances. We drove to the local library where there was a car park and Graham ventured forth in search of instructions. It didn't help, therefore, that the library was closed. After the best part of an hour spent pootling about trying to find out where on earth we were supposed to be going during which I made repeated telephone calls to the vendor of the life-jackets, each time to be greeted by a "number unobtainable" tone.

By this time I was beginning to get cross. This was nobody's fault except perhaps for the hapless vendor who had been stupid enough to have left his telephone unplugged, but these life jackets were absolutely essential and we were already so appallingly over budget that I didn't want to be forced into having to pay £100 for a pair. Eventually Graham managed to find a local who was marginally less half witted than most of his peers, who informed us gaily that the place we were looking for was nearly twelve miles further west. We had been given an address in St Eval and Richard had been led to believe that St Eval was a suburb of Wadebridge.

In fact, as we were soon to discover, it was no such thing.

Outside my colleagues in the CFZ my two best friends in the fortean business are Tim Matthews and Nick Redfern. If you have read this book so far you have already met Tim. It is time now, therefore, for me to introduce you to Nick. I've known Nick, man and boy as it were, for about five years. I first became aware of his existence when I was working as the reviews editor for a long defunct UFO magazine called *Sightings*. As I have stated often in recent years, I am not the slightest bit interested in

UFOs. However, for a few glorious months during 1997 I was hailed by the UFOlogical masses as one of Britain's foremost UFOlogists.

This happened, not because of any groundbreaking achievements in the field, but because I told everyone that I was and they believed me. In my opinion, if you believe in little green men from the planet Zog, you must be a few rizlas short of a spliff anyway, and the fact that the great British UFOlogical movement were so easily fooled into believing that I was not only one of their number, but a luminary to whom they should look up is, I believe, an indictment of said movement.

During my sojourn with *Sightings* I was sent a copy of a rather good book called *A Covert Agenda* by one Nicholas Redfern. Unusually for a book about the subject of unidentified flying whatsits it was actually both interesting and informative. It went through the British government files on UFO related phenomena in the Public Records Office at Kew. I telephoned him for an interview for *Sightings* and over the next year or so we became quite good mates over the telephone. Then in 1998 we met in person at a UFO conference in Southend.

It was an unremarkable conference, but on the Friday night Nick, Nigel (who was my PA at the time and I got dreadfully drunk and behaved very badly. We cut a boozy swathe through Essex society which, four years later is still talked about in places where UFOlogists gather. Three days later I still had a hangover but Nick and I were firm friends.

The following year he came to visit me and Richard in Exeter. As any-one who has read their way through this book so far, or indeed anyone who has ever read any of my other books will know, I don't exactly live an ordinary life. My house is a peculiar mixture of rather dated middle-class respectability and bohemian squalour. I am only too aware that many people find the way that Richard and I (and since early last year) Lisa as well, live our lives somewhat odd and most people find it diffi-cult to fit in to our way of life. However, when they do fit in, Boy do they fit in! I remember when Tony Healey the renowned Australian cryptozoologist came to visit us in 1999.

He came for a weekend and stayed on and off for about three months. Nick was the same. He took to life at the CFZ like a duck to water, and the sight of his angular frame, dressed completely in black with a shaved head became a familiar sight around the suburb of Exwick as he visited us more and more regularly each time accompanied by a burgeoning collection of hardcore punk CDs.

Then in the spring of 2001 it all changed. He visited the United States as a speaker at the International UFO conference at Laughlin, Nevada and when he was there he met a beautiful lady called Dana. Within months they were married and living in Texas, and although we keep in touch by telephone, MSN and email, we haven't seen him now for well over a year.

Nick's main obsession, apart from UFOs and obscure punk records is winding me up about unlikely conspiracy theories. Whenever my computer crashes, or there is an unlikely clunking noise on the telephone line during one of our long rambling late night telephone conversations he blames it on the Bilderbergers or the Trilateral Commission or some other shadowy secret organisation plotting for an overthrow of democracy by the international military-industrial complex. I don't know how much of this nonsense he believes (not much I suspect) but it has become a running joke with us both over the years.

I miss him a lot, but never more than when we finally found the village of St Eval, 'cos I know that unless he had actually been there with us he would never have believed it, and I am certain that when he reads this book he is going to accuse me of making the whole episode up in order to wind him up.

St Eval is quite a sizeable village. In fact as far as villages in that part of north Cornwall are concerned it is probably one of the largest in the area. However (and THIS is strange) it wasn't on ANY of the maps, and all the road signs pointing towards it invariably petered out. It took an hour and a half more before we eventually found it (after almost running out of petrol) for us to find it, and then it was almost by accident. It was as if whoever had built the place wanted it to be as difficult to find as possible.

That part of north Cornwall is one of the bleakest and most unprepossessing parts of the United Kingdom that I have visited. It is open, windswept and fairly featureless except within the narrow wooded valleys that sweep down to the sea. The idea that someone would not only build a sizeable village right in the middle of this bleak wasteland, but then manage quite successfully to keep the place a secret is almost inconceivable but this they managed to do. The village was entirely composed (or at least the bits of it that we saw) of grey prefabricated houses. It was obviously a military base of some sort as there were MoD signs everywhere and the only people that you could see were soldiers or airmen in battle-dress.

Unusually, however, for a small garrison village (and with a brother in HM armed forces I have seen several of these over the years), many of these soldiers and airmen were armed. I have recounted elsewhere some of the weird stories that are floating around the westcountry fortean community regarding mysterious occurrences in and about various MoD establishments in north Cornwall especially the vicinity of the listening post at Morwenstow. But I have never come across anything like this.

Armed men. A secret garrison. Strange radar dishes surrounded by barbed wire, and even - once - the sound of distant gunfire. The whole place was like a set from a mid 1970s episode of Dr Who and was probably the weirdest and most unpleasant place I have ever been in England. During the hour or so that we spent driving around the village searching for the home of the guy who was supposed to be selling us a pair of life jackets I was acutely aware of being watched and felt that if we were to put even one foot wrong we were likely to be immediately arrested under some obscure statute which has never heard of *haebeus corpus*

During my chequered career as a fortean investigator, and indeed in other spheres of my life, I have been in some unusual situations. However, in many ways I have never felt so uneasy as I did in that creepy little village. I looked at Richard. Richard looked at me and said:

This is gonna end up in your book, isn't it?"

I nodded agreement and fervently wished that I hadn't given up smoking. There really wasn't anything left that either of us could say about the matter.

More by luck than by judgement we found the place where the bloke who was selling the life jackets lived. Richard and Graham knocked on his door, gave him thirty quid, took the life jackets and within a minute we were doing our best to find the quickest route of egress from the camp. We drove past a school. Even that had a sinister looking group of uniformed men standing guard as the children were playing a desultory game of football on the dry and unhealthy looking grass. In the middle distance was a long row of low, grey concrete buildings. From the distance that we saw them, they didn't appear to have windows and appeared to be surrounded by a barbed wire fence. There were soldiers patrolling up and down the fence. Whatever it was inside the bunkers had to be worth guarding.

"If bloody Redfern or his Taffy mate were here"... said Richard (referring to notorious fence jumper and base invader Matthew Williams), *"they would be hypothesising that there was a crashed UFO in there..."*

"Frankly I don't care what there is in there" I replied *"let's get the hell out of here."*

So we got the hell out of there.

An hour and a bit later we were in Plymouth where we were to meet our old friend Chris Moiser. Chris is a teacher at Plymouth College of Further Education, and he has been a core member of the CFZ team ever since - in the summer of 1995b- he was the only person to turn up at a public lecture that I had been booked by Exeter Education Authority to give on the subject of mystery animals.. He is a kind and gentle geezer a few years older than me and his quiet good humour and gentle common sense has made him an invaluable member of the CFZ team. Both his zoological acumen and his knowledge of the law have been invaluable to us on a number of occasions and have got us out of a lot of scrapes.

We were supposed to meet Chris at Plymouth College itself, but we were ejected by the security guards who didn't like the look of us for some reason, so we drove down the hill towards Devonport until we saw him waving cheerily at us from the side of the road. He was clutching a black leather case which contained a Russian made nightsight which we were going to borrow to use on our forthcoming adventure.

Chris navigated us to a quiet backstreet pub where we settled down for the evening. Chris demonstrated the use of the nightsight to Graham, while Richard and Lisa bought a round of drinks, and I telephoned my friend and boss Simon Wolstencroft. Simon is the editor of *Tropical Fish* Magazine who I write for on a monthly basis, and is directly responsible therefore for financing some of the CFZ's activities. He is also a party animal *par excellence* and so when he arrived at the pub it was a fore-gone conclusion that we would be there until closing time.

After we had been there about an hour Graham sloped off to play a game of pool. He is a complete addict of the game and plays it wherever and whenever he can. In *Only fools and Goatsuckers* I tell the story of how he even ended up playing pool with the Civil Defence team in the mountains of Puerto Rico when they could hardly speak English and Graham could hardly speak Spanish. That night he and a swarthy looking bloke called Ernesto ended up teaching each other trick shots and a splendid time was had by all. Nothing as exciting happened this night in Plymouth. Graham continued to play pool, and eventually Lisa joined him, while Chris, Richard, Simon and I drank, swapped zoological anecdotes and generally enjoyed ourselves until the barman rang the bell for 'last orders'.

We said goodnight to Chris, poured Simon into the Jag and after dropping him home drove back to Exeter. The next day, Graham carried out a series of routine tests on the equipment. In the afternoon we did a photo session for a photographer from the *Exeter Express and Echo* - for the first time showing John, Graham and Richard resplendent in the new CFZ field kit or uniform, and then we found that there was no time left for preparations. It was time to embark on the second stage of our investigation into the Monster of the Mere.

158

Memo

Re:CFZ's lake contour mapping and search for fishes, 25th - 30th July 2002

STATEMENT OF METHOD

Phase 1: Arrival, general preparations, familiarise ourselves with the site.

Phase 2: commencing on day 2

Aim: The underwater contours of Martin Mere are to be determined by sonar measurements taken from a surface craft - an inflatable dingy. The lake will be scanned in strips, one strip at a time.

Method: The strip will be physically delineated by means of string or lightweight rope stretched across the lake and secured at each end by a peg hammered into the ground. An operative will be in the vicinity of each peg - for public safety reasons; to adjust rope tautness as required; and to move the peg and rope sideways when it's time to start a new strip of the lake.

The dingy, crewed by two further operatives, will make its way along the rope, and sonar readings will be taken at regular inter-vals - probably one-meter intervals - and the data duly recorded. One operative will propel the dingy whilst the other operates the sonar and the dinghy's handheld walkie-talkie.

Phase 3, to commence when phase 2 is concluded:

Aim: explore the lake - mainly its deeper channels - for a "lake monster".

Methods:

* Use of sonar fish finder: We intend to criss-cross the mere in a dingy scanning for the fish. The device reads water depth and shows up any fish present.
* Baiting smaller fish: We will lay bait in one area to attract smaller prey species in the hope that they in turn will attract the "monster"
* Fish chumming: We will drag sacks of bait across the bed of the mere in the hope that our fish will be attracted. If this works we intend to lure it to the surface and photograph it.
* Use of night-sights: We will watch the mere in shifts after dark with a night-sight to see if the fish surfaces.
* Witness interviews: We will speak to witnesses about their sight-ings, taking note of the time and area of the mere they occurred in.

Notes:

* At the end of each working day, all equipment is to be removed and safely stored in an appropriate location.
* Jonathan Downes will be in overall command of the operations and in radio contact with the other operatives when appropriate.

159

PRESS RELEASE

****** For immediate release ******

Date of issue: 18/07/02

Experts try and uncover the mystery of the Mere

An investigation team from the Centre for Fortean Zoology are spending four days at WWT Martin Mere, Burscough, Nr. Ormskirk, Lancashire from Thursday 25th July to try and uncover the cause of the mysterious underwater activity witnessed by visitors to the mere over the past five years.

Most recently, six months ago, several visitors were left shaken up after witnessing a Swan trying to flee the grasp of a giant under-water predator. Other witnesses, including the Centre Manager, have witnessed an extremely large creature in the depths of the lake.

The four man team from The Centre for Fortean Zoology, based in Exeter, plan to spend four days watching over the mere both day and night, utilising infra-red cameras, military style night-sights, and 'fish finder' sonar equipment to try and uncover what is thought to be a giant Wels Catfish "If it is a wels" says CFZ Director Jonathan Downes "It is almost certainly a British record".

The activities have become a major talking point amongst locals of the villages that surround WWT Martin Mere, one of nine Centres operated by the Wildfowl & Wetlands Trust (Founded in 1946 by the late Sir Peter Scott).

ENDS -

For further information please contact:
Carl Lamb, Marketing Manager
Pat Wisniewski, Centre Manager
Jonathan Downes, Director, The Centre for Fortean Zoology

NB From Thursday 25th July Jonathan Downes and the other members of the CFZ Team will be available on his mobile.

Note to editors:

WWT is the only registered wildlife charity in the UK dedicated to the conservation of wetland birds and their habitats worldwide. Further details about the WWT can be obtained from http://www.wwt.org.uk/

The Centre for Fortean Zoology (founded 1992) is a registered non-profitmaking trust. It is the only organisation in the world dedicated to the professional and scientific investigation of unknown and mystery animal species

Further details about The Centre for Fortean Zoology can be obtained from http://www.eclipse.co.uk/cfz

Background Information

*   Wels catfish are native to mainland Europe and were introduced into parts of the UK in the late 19th Century. They are the largest freshwater fish in the world, can reach a length of 16 feet, and have even been reported to kill and eat people.,

*   Martin Mere is the remnants of what was once England's largest lake. It was once five miles long and surrounded by treacherous bogs and marshes. Over the last 800 years the swamp has been drained leaving only a modest lake. But Martin Mere has not surrendered all its secrets yet.

Jonathan Downes and Richard Freeman, who spent a week investigating these sightings earlier this year, are leading a four man expedition to probe the Mere's opaque depths. Cryptozoology is the branch of zoology that deals with unknown animals.

They are the only professional cryptozoologists in the UK. They have pursued vampires in Mexico, dragons in Thailand, and skunk apes in Florida. Closer to home, in the UK, they have hunted big cats, sea serpents and the grotesque Cornish Owlman. The CFZ team are hoping that The "Monster of the Mere" will not be the ultimate fisherman's tale of 'the one that got away'.

PUBLICITY LEAFLET

Following a week spent at Martin Mere earlier in the year by CFZ Director Jon Downes and Cryptozoologist Richard Freeman, a four man team will be at Martin Mere from 25th-29th July. With full co-operation from the staff and management at the Wildfowl and Wetlands Trust Reserve, they will be investigating reports of a giant aquatic predator which is attacking waterfowl as big as swans on the lake. They will be utilising not only their zoological knowledge but an impressive array of hi-tech equipment in their search for what they believe is probably the largest wild freshwater fish in Britain.

ACT TWO

"Everything you can think of is true,
The baby's asleep in the shoe,
Your teeth are buildings with yellow doors,
Your eyes are fish on a creamy shore…

Tom Waits and Kathleen Brennan

The carved dragon on the old house.
Could *this* be the *Orm* of Ormskirk?

Richard's and my first sight of the Mere. It looked bleak and uninviting
and we felt sure that there could be something nasty in there!

Blackpool; possibly the nastiest place I have ever been

St Eval—I only wish Nick Redfern had been there to see it

Blackpool Tower; in the foreground hordes of the great unwashed
consumer carcinogenic rubbish

Two views of the Wintergardens complex in Southport at the end of the 19th Century. The top picture shows the opera house, the lower the concert hall. The aquarium was situated beneath the two.

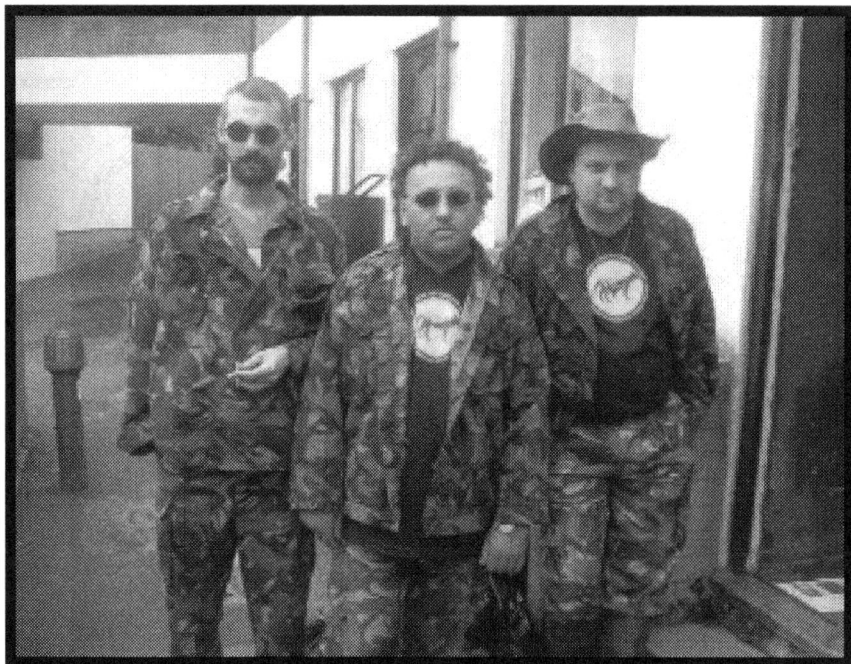

The military wing of the Centre for Fortean Zoology

Testing the *Waterhorse* in the carpark for the benefot of
the Exeter *Express and Echo*

Experts close in on 'Mere Monster'

A giant fish which has attacked swans at a bird sanctuary has been spotted by wildlife experts.

At least two swans have been hurt by the underwater creature nicknamed the Monster of Martin Mere which hides in a lake in West Lancashire.

Now a four-man team says it has located the attacker which they think could be a Wels Catfish from eastern Europe.

Although the biggest ever caught was 16-feet long, Jonathan Downes, one of the team from the Centre for Fortean Zoology in Exeter, thinks this one is a tiddler by comparison - perhaps seven feet long and weighing 24 stone.

He said: "If it is a Wels, it is almost certainly a British record."

The team is spending four days at the Wildfowl Trust at Martin Mere using infra-red cameras, military-style night lights and sonar equipment to find out more about the mystery beast.

The giant fish was first spotted on Thursday by team member Richard Freeman who is hoping to capture the creature on film.

He said: "I have seen something black and shiny snaking around in the water in almost the same place as the original sighting several months ago.

"It certainly looked like a Wels catfish.

'Rubbery' body

"However we will be carrying out further investigations over the weekend in hope of obtaining photographic proof".

Mr Freeman said the fish had no scales, had a "rubbery" appearance, was oily-black in colour and moved quickly through the water.

"I can't say for sure that it was a Wels catfish. But if a pike had attacked the swans there would have been wounds.

"This thing seems to come up underneath and drag its prey down under the water."

Reports of a larger-than-life creature living in the 17-acre lake were first voiced four years ago and the Martin Mere monster has since become a talking point among people living near the 380-acre reserve which regularly attracts Whooper and Bewick swans.

The mere is home to scores of swans

OPERATION MARTIN MERE causes a stir with the media

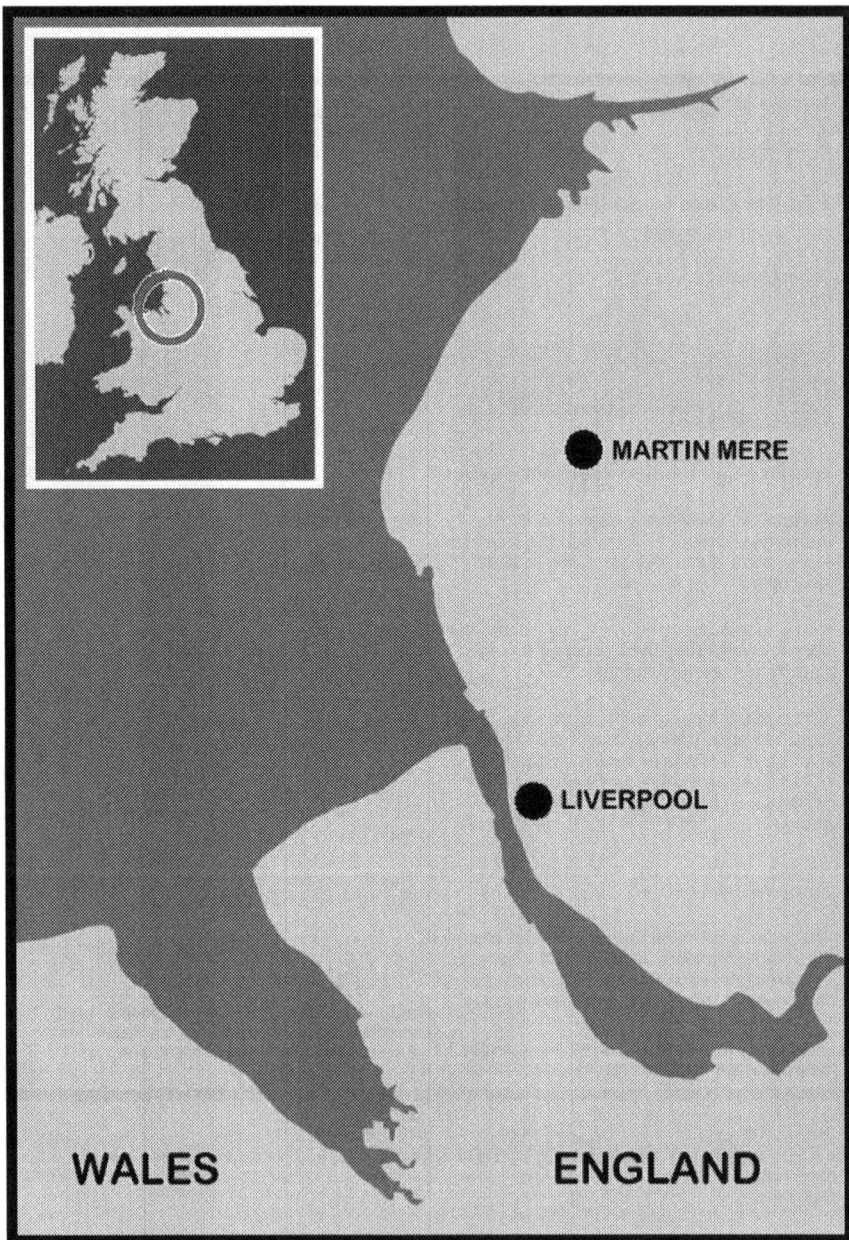

(This Page): Martin Mere in relation to the rest of the country.
(Opposite): This was the best map that we could find of the nature reserve. However when you compare it with Graham's map, you will see that it is woefully inadequate. (Courtesy WWT)

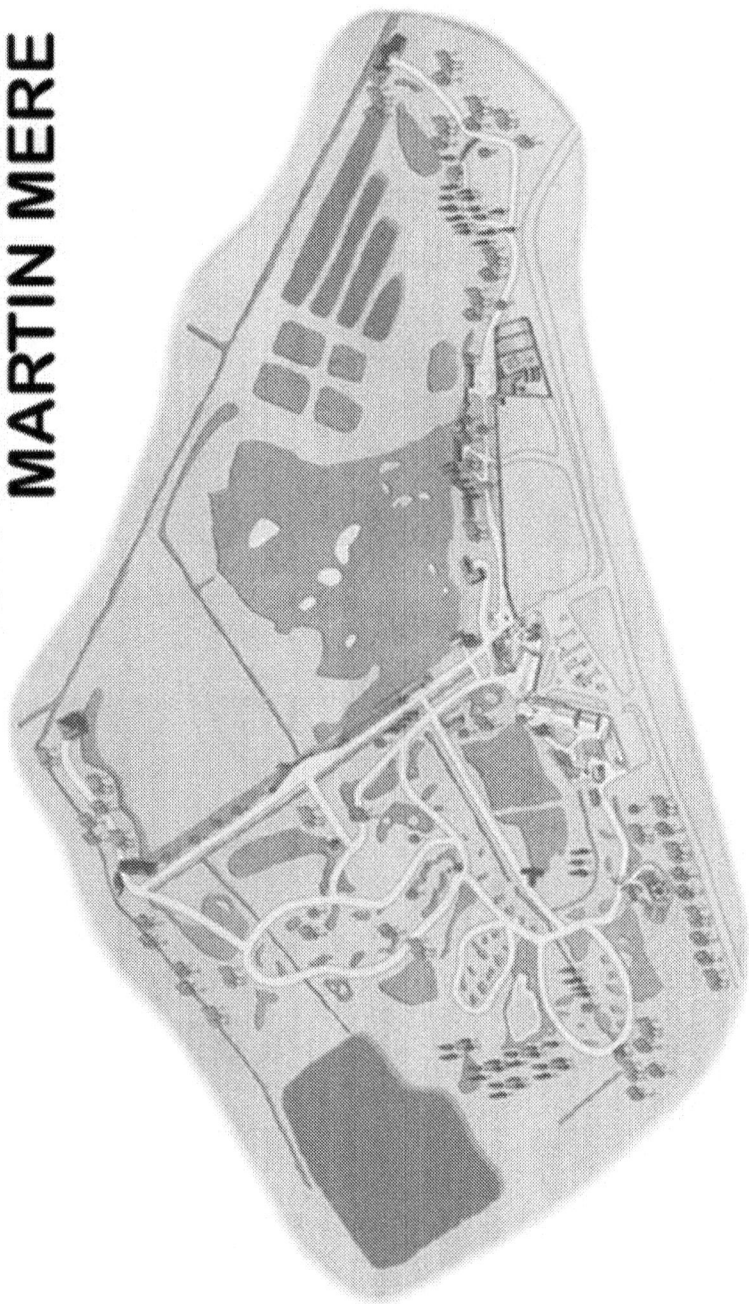

WWT   The Wildfowl & Wetlands Trust
MARTIN MERE

The Mere at dusk. Dark and mysterious.
A place where *"nothin' would surprise"*...

'Rebecca's Channel' - between 'Biro Island' and the shore. Was *this* where the

The far side of the lake

Fish Base 1—l-r Richard Freeman; John Fuller

Graham lies on his stomach in the *Waterhorse* as he gingerly pilots her through
`Rebecca's Channel`

L-R Alexandra Matthews' Tim Matthews;
the author; Richard Freeman; John Fuller

The author at his laptop in the Nerw Raines Observatory

The injured swan at the far side of the lake

The Daisy Causeway

Frank Buckland 1826-1880

What a whopper! These two anglers struggling to
hold up this giant Wels catfish (*Silurus glanis*)

THE AQUATIC INVADERS

Chinese mitten crab
How exotic species are changing our rivers and lakes

Probably carried here by ships and now common in the Thames, Humber and other rivers. Its deep burrows undermine river banks, leading to collapse.

INTRODUCED SPECIES

Beluga sturgeon
Commercial fisheries want to import sturgeon because it fights so well when caught and because it can grow to 1.25 tons

Red-eared terrapin
Aggressive reptile which can grow to the size of a dinner plate and inflict a nasty bite. Experts say it is wiping out the endangered great crested newt

ENDANGERED OR EXTINCT

Burbot
Once common across East Anglia, Yorkshire and the Trent, this miniature catfish-like creature has not been seen for more than a decade and is probably extinct

Pumpkinseed
Imported from the USA as an ornamental fish but now common in rivers across southern England. Believed to eat the eggs of native species

Vendace
Once common across Britain but now restricted to Bassenthwaite and Derwentwater in the Lake District where it is under growing threat from the roffe, a small perch-like fish which eats its eggs

Wels catfish
Reaching more than 150lb, and renowned for catching waterfowl and swallowing them whole. Increasingly popular with anglers who have introduced it to waters across the Midlands

Graphic by Roy Cooper & Julian Osbaldstone

Anglers' imported monsters endanger native fish stocks

by Jonathan Leake

WHEN a five-year-old girl recently caught a young sturgeon in a boating pond in Basildon, Essex, it was a symptom not of a nascent local caviare industry but of a craze that is threatening Britain's native freshwater fish.

Carp weighing more than nine stone, giant catfish that can swallow a waterfowl and huge sturgeon are among species introduced to Britain's waterways and lakes to meet anglers' demands for bigger and more thrilling catches.

Experts fear the imports could introduce new diseases, diminish food stocks and kill off native fish. Almost a fifth of Britain's 55 breeding species are now considered to be endangered. Two, the burbot and the bouting, have recently been declared extinct.

Dr Bruno Broughton, a fisheries consultant from Not-tingham, blames the growing numbers of commercial fisheries for anglers. "Owners run these fisheries to make a profit and if they put in lots of big aggressive fish they get more customers. The trouble comes when those fish escape," Broughton said.

Robin Goforth, owner of Hayfield Lodge fishery near Doncaster in Yorkshire, imported 200 young beluga sturgeon, which can grow to more than a ton. He is fighting an order from the National Rivers Authority (NRA) to send them back. "Beluga are one of the largest fish in the world and I thought they would be great sport," he said.

There was similar controversy when another commercial fishery imported two massive Chinese blue carp. An angler who caught one weighing 52lbs claimed the record for Britain's heaviest coarse fish. It triggered a flood of similar imports, which infected British carp with SVC, a viral disease that can kill all the fish in a lake.

Mick Toomer, who runs five fisheries in Essex, said customers often asked him to introduce exotic species. "Some people are importing catfish which are larger than the British record so that their customers can become record breakers," he said.

The introduction of exotic fish has been going on for centuries. At least 17 new breeds have been put into Britain's waterways and 12 have established sizeable populations. The rainbow trout, brought from America because it is easy to catch, has driven out the native brown trout in many southern rivers.

Malcolm Elliott, head of fish biology at the government-funded Institute of Freshwater Ecology, wants tighter controls. He said: "New species may compete for food, bring disease and parasites or simply eat the native fish."

Not all foreign species are, however, brought in by anglers. Chinese mitten crabs are thought to have arrived in the ballast tanks of ships. Their burrows are undermining sea defences in the Thames and other British rivers. Other waters are infested with red-eared terrapins, bought as pets during the Ninja Turtles craze but released when they grew into large, vicious adults.

Article taken from The Sunday Times, 17th December 1995

Thursday 25th July

There was an episode in the first series of the seminal BBC TV comedy series *The Young Ones* where one of the characters was seen to stagger out of bed and yawn. He then said something to the effect of *"Crikey Lads, it's two o'clock in the afternoon"* and the audience exploded into laughter. Why everyone finds the idea of people staying in bed until that time of the day highly amusing I don't really know, because I have always been semi-nocturnal and I am usually awake, working until the small hours and often do not rise until lunchtime and sometimes later. It is one of the things that the three members of the core CFZ team have in common. None of us are early risers and so for all of us to be up, dressed, and relatively fresh at five in the morning usually means that we have been up partying for most of the night before. However on this occasion it meant something completely different. For the first time since Graham's and my highly contrived visit to the Y100 Radio studios in Fort Lauderdale during our Chupacabra adventure (see *Only fools and Goatsuckers* for details) the CFZ were up and on parade at a disgusting hour of the morning. Reasonably fresh and alert and ready for whatever the days ahead had to throw at us.

Most of the gear had been packed the night before and I think that it is testament to the growing efficiency of the rag-tag and bobtail organisation that is the only full time cryptozoological research team in the world, that when we got to our final destination the only thing we had

181

forgotten to bring with us was a single wooden stool - which was hardly the most important item of equipment.

By twenty past five we were ready to go and, still in the semi darkness, we drove in convoy off down the road towards the local Sainsbury's superstore where we were to fill up with petrol. It may not be as romantic as the accounts of the great historical explorers like Mungo Park and Percy Fawcett, but such things are an unfortunate necessity of life. Choosing to turn a deaf ear to Graham's comments that Sainsbury's petrol stations were serviced by one of the less eco-friendly multinationals, I decided that as I not only had a Sainsbury's reward card but discount vouchers for petrol, eco-warrior concerns had, for once, to be buried beneath a tide of pragmatism.

Books which are essentially tales of expeditions and high adventure don't usually mention the more prosaic side of expedition life. However, when one starts to think about it there is an enormous amount of "stuff" that is necessary when even a modest sized expedition like ours takes to the road. The two cars were therefore loaded to the gunwhales, not only with the exciting things like the boat, the fishfinder and the radio equipment but with a couple of hundred quids worth of canned food, bog paper, wet wipes, camping gear and, of course, cans of lager.

A few years ago I had a spat with a Norwegian gentleman called Jan Ove Sundburg. He had heard tell of various late night drunken shenanigans that had taken place at the Fortean Times Unconvention and from that had drawn the inference that I was dead drunk all the time I was engaged in field work. This was a totally unwarranted slur and sparked of a slanging match on the Internet between him and both me and an Irish guy called Daev Walsh who had been unfortunate enough to haven accompanied Sundburg on one of his expeditions. Daev had left the expedition (to a Swedish lake reputedly the haunt of a monster with the same name as one of Homer Simpson's sisters in law) after only a few days, and ended up being accused of being a Devil worshipper whilst I (who have never met the geezer) was accused of being a raving alcoholic, and eventually "a disgrace to my country". I threatened to punch Sundburg on the nose if I ever saw him face to face, and after a flurry of vitriolic emails which eventually led to him being banned from at least one on-line community, the matter rested.

The previous evening as we had loaded up the car with our voluminous amounts of groceries I smiled at myself at the thought of Sundburg as we loaded a couple of trays of lager into the boot of the Jaguar. *"Screw Him"* I thought happily, and continued preparations for our big adventure.

As I sit typing this on my laptop, I am in a bird hide at Martin Mere itself, sitting looking over the water at Graham and Richard posing in the rubber dinghy for a photographer from the *Liverpool Echo.* I have not got any of my enormous and ever expanding library with me, and for once I haven't even brought along the trusty copy of *Moby Dick* which usually accompanies me on our adventures. It is, after all, the greatest literary excursion into the world of monster hunting that has ever been written.

There is a classic line in Mellville's book which reads something like, 'we sailed intrepidly into the dark Atlantic'. I thought of that passage and grinned to myself as our little flotilla of cars; my father's old blue Jaguar leading the way and the battered old CFZ Volvo Estate bringing up the rear as we drove intrepidly into the dark M5.

We had a long drive of nearly 270 miles ahead of us. We had decided to set off early so we would avoid being anywhere near Bristol, the West Midlands Birmingham/Walsall/Wolverhampton conurbation, or Manchester and its satellite towns during the rush hour. Because both our cars, and especially the Volvo are somewhat elderly we decided, whilst not exactly driving up in convoy, to break the long journey at three pre-arranged stop-off points, at one of which we would eat breakfast. The first of these stops was to be the big motorway services at Easton-in-Gordano outside Bristol. Both Graham and I are so familiar with the stretch of road between Bristol and Exeter that we could probably have driven it in our sleep, and so we drove this section of the journey almost on autopilot. For the first quarter of an hour or so we drove along the dark road we played with the walkie talkies. Richard acted as my 'Radio Officer' and John did the same for Graham. The novelty of pretending to be fighter pilots wore off after a while and as Richard dozed off and snored loudly I switched on a classical music station on the car radio. It was playing a selection from Puccini, which pleased me greatly, and so I hummed along to *La Boheme* happily as I drove and daydreamed about my burgeoning romance with Becky.

The Internet hasn't been around long enough for people to truly accept

the impact that it has, and more so, that it is going to have on the machi-
nations of society. Several years ago after reading Bruce Sterling's *The
Hacker Crackdown* and Julian Dibble's *My Tiny Life* I wrote a long and
reasonably tortuous essay about, what I dubbed, as cybersociology. In
five year that I have been online, I have joined, left, started and closed
several cyber-communities. I have indulged in cybersex and have experienced, either at first hand, or vicariously, or by reputation much of the
gamut of human experiences that are available in cyberspace. Dibble described such experiences as *tiny*experiences. He refers to *tiny*friendships,
*tiny*communities, and even *tiny*sex. In the five years since Dibble wrote
his book, the complexity of online sociology has increased manyfold.
Most people under the age of fifty now know at least one couple who
met through the Internet. I know several, one of them being my now de-
ceased next door neighbour, and the complexities of what is vulgarly
known as the "dating game" have changed forever. I must admit that I
have always considered the idea of cybersex to be an amusing diversion
rather than anything serious and concrete. My forays into cybersex were
fun but had no deep emotional significance, and I had pretty well given
them up. However everything changed when, late one evening sometime
between our first and second forays to Martin Mere, I wandered into a
chat room on MSN where, I had been told, there was an online trivia
game.

I have always been fond of pointless general knowledge quizzes, and so,
being bored and mildly stoned, I decided to play. Whilst I was there I got
chatting to a woman from Maryland. Her name was Rebecca and she
was a year or two younger than me. We chatted online for anything up to
six hours a night, each night, for the next week or so, and then - partly by
mutual consent and partly by accident - in our respective bedrooms thousands of miles and an ocean apart we seduced each other with words.
Within days, to use Dibble's terminology, I was in *tiny*love.

I didn't know then, and I don't know now whether our relationship has
any future outside of the Internet, but I now know, that no matter how
ludicrous it sounds, it *is* possible to have strong and passionate feelings
for someone whom you have never met IRL, and to whom you have only
spoken on the telephone twice. Furthermore, in the history of my chequered and fairly unsuccessful love life I have never yet been involved
with a woman who will stay up all night just to be able to wish me good
bye and good luck as I embark on an expedition. My ex-wife wouldn't
have done that, and neither would any of my other girlfriends (who have

184

ranged from the slightly detached to the criminally insane), but this is exactly what Becky did.

Despite having to get up for work at some unearthly hour before 6 am, she waited up until after 2 am (her time) in order to wish us *bon voyage.* As I told her that morning (again with Herman Melville on my mind) through the miracle that is MSN Messenger, she made me feel like, the captain of a 19th Century whaler, being bade farewell by his grieving wife, who was putting on a brave face although she knew that her lover might never return from the sea. All this might seem a trifle melodramatic as a quasi-literary simile for someone who essentially is just going to Lancashire for four or five days, but the expedition had taken so much preparation and had cost so much to set up, and indeed had taken up so much of my mental and physical energy since it was first mooted at the end of February that it had become to assume the status of a semi mythical adventure of epic proportions within my manic-depressive psyche.

Glowing from the thought that for once I had a woman who cared enough about me to wish me farewell and who would be waiting for me when I came back (even if I hadn't actually met her IRL) I drove on towards Bristol. I don't actually know when the M5 was first built but I do know that when I first visited Devon during the long hot summer of 1969, it certainly didn't reach as far as it does now. It is a spectacularly unattractive piece of road, but the sun came up as we drove across the Somerset levels, and the first rays of the new rising sun lit up the feathery gossamer on the willow trees which lined the waterways with gold, and for a few brief moments, as *La Boheme* reached an orgasmic climax and Richard snored away dreaming of disgusting sexual exploits with gothic amazon women, I was transported to fairyland.

Et in Arcadia ego. Yes, I too once dwelled in Arcady, but Jesus H Christ, I never expected to find it at six thirty in the morning along a dilapidated stretch of the M5.

We drove on steadily northwards and as I could see the sprawling mass of Bristol before us as we approached it at about 60 mph, Richard woke up and yawned. Just then, a message came through on the walkie talkie from John. Clearly we were back within radio range. Apparently they had been waiting in the car park at the Gordano services for about a quarter of an hour. A few minutes later we joined them and we took a break of about a quarter of an hour. We took advantage of the washroom facilities, purchased bars of chocolate and drove north again.

There is only so much gripping prose that one can write about the joys and perils of motorway driving. The joys are few and far between and the high spot of the journey up was undoubtedly the sun rising over the Somerset levels. The perils we will come to anon, and apart from them there is really very little to say about the next leg of our journey which took us up to an undistinguished and rather ugly motorway service station just north of Bromsgrove. Now, I know very little about Bromsgrove apart from the fact that it is grubby, ugly and flat and that I lost my virginity to a girl who hailed from there more years ago than I care to remember and so the chances of me waxing lyrical about the place are absolutely nil! What I can say is that we were all getting rather hungry by now so I dug into my ever lightening wallet and splashed out thirty five quid on four overcooked "full English breakfasts" which tasted like damp cardboard and probably clogged up more of my arteries than I would like to admit. However the intake of protein, caffeine and stodge did us all the world of good and after a rest of about an hour we resumed our journey northwards, agreeing to meet up at another motorway service station at Knuttsford, just outside Manchester.

Now I ain't a fast driver. My reactions are quick enough to make me a perfectly competent one, but not if I am careering up the motorway at a speed of greased lightning, so I usually poddle along at between 45 and 65 mph. Various friends of mine are appalled that when I am driving a high powered, high performance vehicle like the Jag, I keep to the speed that one of my friends described as being like some old biddy from Cheltenham Spa, and another equated to that of a three toes sloth on mandrax, but I keep to the speed at which I feel comfortable, and unlike most of my friends and acquaintances I have never yet (touch wood) been busted for a speeding offence.

I am quite aware that people who drive too slowly on a motorway are both dangerous and an utter bloody nuisance. On occasions I have been as guilty as anyone else of getting irritated with the sort of person who drives up a motorway at a snails-pace of 25-30 mph. However, to my mind at least an average speed of just under 60mph is a perfectly acceptable speed in the slow lane of the M6. Unfortunately a large number of Belgian lorry drivers do not appear to agree with me.

I dunno what it is about the Belgians. Maybe they are merely a nation inclined to breed ill mannered lorry drivers. Maybe it was just my imagination. Or maybe the dark forces of the Belgian secret service had taken exception to an article I wrote years ago berating the Belgian Colonial Service's attitude towards both animal and human rights in what was then the Belgian Congo, and had dispatched a crack team of their best espionage juggernaut drivers to settle the score with me. However, for a stretch of about seventy miles up the M6 what seemed to be an endless procession of Belgian lorry drivers, each driving vehicles bearing the distinctive logo of a well known pan-European haulage company did their best to drive me insane.

Although I didn't seem to have aroused the ire of any of the truckers of other nationalities on the M6 that day the Belgians seemed to be mightily pissed off with me for some entirely unknown reason. The enormous lorries would creep up on me, and then when they were just behind me flash their lights pugnaciously and blow their air horns, in some cases performing an internationally recognisable gesture of sexual contempt in my direction and gibbering in Walloon.

When you are a manic-depressive with strict instructions to avoid stress, being systematically bullied by a seemingly endless parade of muscles from Brussels is not exactly what the proverbial doctor ordered. The journey between Brum and Knuttsford seemed interminable. It was like some exquisite form of Roman torture, or an idiot Nintendo computer game whereby a parade of Belgian lorry drivers try to either drive a fat cryptozoologist insane or force him into a ditch. This continual harassment only served to make me drive even more cautiously which in turn aroused the ire of the Belgian truckers even more. By the time we finally arrived in Knuttsford I was exhausted and ready to drop.

We pulled into the services ready to meet our colleagues and quaff a long awaited and well deserved cup of tea. But it wasn't to be.

Fate had other plans in store for us, because although we drove round and around the car park looking for Graham and John they were nowhere to be seen. I rang Graham's mobile to be informed cheerfully that they were 'somewhere north of Wigan'. *"Bloody hell"* I said. *"That means that you are something like 40 miles further north than us, and if you are not careful you will end up in Preston or Blackpool".*

Using John to relay my navigation instructions to Graham we managed to get him off the motorway and onto the little B road that led to Burscough. Forgoing out cup of tea we rejoined the M6 and drove as fast as I could manage northwards. Even when I was approaching the national speed limit the serried ranks of Belgian lorry drivers continued to harass me over the next forty miles or so, and I made a mental note to lampoon the entire country of Belgium as brutally and as unkindly as I could if I ever get around to writing another novel.

Eventually we left the motorway and took the same B road that our companions had done twenty minutes or so before and drove towards the little canal side town of Burscough. Since I was a small boy I have always had an eye for familiar landscapes, and as Richard and I drove along the roads that we had explored a few months earlier, in many cases it was like revisiting old friends. THERE was the garage where the imbecilic girl had told me that my credit card was invalid. THERE was the shop where we had bought our lunch one day. And THERE was the roadside lay-by where we parked and had lunch after Lynda had told us that she was going into labour.

What made this leg of the journey even more poignant was that in the seven weeks or so that had elapsed since our first visit to the area we had carried out an enormous amount of research into the history of the place, and especially after having transcribed the two documents that are printed in this present volume as appendices names like Scarisbrick, Lathom and the like had assumed a new and fascinating significance for both of us.

As we drove along happily we noted the wood where the child had been found ensconced in an eagle's nest. We saw the stretches of the canal and sluices that had been paid for over the years by various scions of the Scarisbrick family, and many other local landmarks besides. Amidst a helter-skelter of "Do you remember's" from both of us, we hardly noticed that we had arrived in Burscough itself and were ready to rendezvous with our errant companions.

Over the mobile telephones we had given vague instructions to Graham as how to find his way to the library car park where we had parked during our ill fated trip to Burscough in late May. Because my half remembered instructions had been so vague I wasn't really expecting that they would have actually found the place and so we had decided to drive there ourselves and park before telephoning the others, ascertaining their location, and doing our best to meet up with them.

However as we drove into the car park there they were. Black John was wandering around looking cheerful and obviously doing his best to grok as much of west Lancashire as he could in the shortest stretch of time, and slightly to my consternation, Graham had the bonnet of the Volvo up and was fiddling around beneath it with a worried look on his face. We drove in, parked, disembarked and hailed them. As it turned out there was absolutely no need for me to have worried. Graham was merely trying to adjust the settings of the windscreen washer in an idle moment while they were waiting for us.

We greeted each other cheerfully and talked about this and that for a few minutes as I hobbled around on my stick and tried my best to get some circulation back in my nether regions. Giving that up as a bad job I hobbled back to the car and we drove in convoy to Martin Mere.

The tiny convoy drove down the narrow straight road towards the Mere and turned into the well remembered car park. The expeditionary force of the Centre for Fortean Zoology arrived at WWT Martin Mere. We didn't look much like either a reputable scientific expedition or indeed a quasi military group of well disciplined men. The two vehicles looked appalling. The Volvo (which had cost me a hundred quid back in April) began to look as if we had been robbed. My beautiful Jaguar was covered in a thin film of Devonian alluvium. The fields on the hillside on the opposite side of the valley to my house had been raped and pillaged by the massed JCBs of the property developers. They had destroyed the flowery fields where I used to roam with my dear old dog Toby back in the days before he became a senile old hound and I became a semi cripple. Now it was ground to dust in order to provide badly built and characterless homes for a new generation of people that I had no interest in getting to know unless they could be persuaded by any remarkable sleight of hand to purchase any of my books and records. However the aforesaid dust had covered everything in the valley with a thin film of filth. It wasn't just the cars that looked unkempt. Graham and John looked quite dapper, but Richard was wearing black trousers and a rather faded T Shirt advertising an eccentric indie band called *Half Man Half Biscuit* and I was wearing a pair of shorts which, according to Richard at least, made me look like a far and boorish American tourist and, when I caught sight of myself reflected in the car window, reminded me of P.G. Wodehouse's description of someone who made one re-evaluate the concept of man as the pinnacle of God's creation. We really should have spruced ourselves up before arrival, or maybe even donned our CFZ uniforms but we didn't and the die was cast. We marched sheepishly up to the entrance kiosk where the two pretty girls who had greeted Richard and me so warmly six weeks before flashed welcoming smiles at us, recognising us immediately and ushered us into the visitor's centre itself.

We asked to speak either to Pat or to Carl Lamb, the marketing director, and with even broader smiles we were told to sit and make ourselves comfortable and that Carl would be there in a few minutes to meet us. As it was it was only about thirty seconds later when Carl, an affable young man, whom it turned out, had an even more vile sense of scatological humour than any of the CFZ bounded across the foyer to meet us. After mutual introductions and welcoming handshakes all round he ushered us into the cafeteria and provided copious amounts of tea and coffee.

Even before I had met Carl, it became obvious during our long and convoluted telephone conversations during which we had discussed the best way to glean publicity out of our expedition, it became obvious that Carl was one of these people who is constantly doing about twenty five things at once. At the very moment that he sat down with us in the Martin Mere cafeteria to drink tea and discuss our planned *modus operandi* for the following four days, his mobile phone rang and he scurried off in search of an answer to whatever query he had been confronted with.

Over the next four days this became a familiar pattern. Carl couldn't have been kinder and he managed to engender an enormous amount of publicity for our endeavours over the next few days. However, every time one would sit down and try and have a conversation with the man, it would be interrupted by the familiar sound of his mobile playing the theme song from *The Simpsons* (something which immediately endeared him to both Richard and me who are confirmed Homerholics) and he would rush off somewhere like an earnest and extremely efficient young dragonfly, never to be seen again. Well not for an hour or two anyway.

As we sat down drinking our tea and lazily looking at the Hawaiian geese which rootled about in the earthy bank outside the great picture window of the cafeteria, Pat came striding in to meet us. He had a harassed look on his face. Now, as anyone who knows me will be only too aware, I do have a distressing tendency towards paranoia in my makeup. The previous afternoon back in Exeter I had telephoned Pat about some trifling matter or other and he hadn't sounded as affable and as enthusi-astic as he usually did. My paranoia circuits immediately went into over-load and I began to imagine a whole selection of bewilderingly unlikely worst case scenarios.

* MAYBE he had been contacted by Larry O'Hara. O'Hara had done his best to banjax Tim's career over the years, and as I was now a reasona-bly well known friend and associate of Herr Matthews (him and Lynda had recently become the Lancashire representatives for the CFZ), was it possible that O'Hara had contacted Pat claiming that I was guilty of some appalling neo-Nazi hate crime?

- MAYBE he had managed to get hold of a copy of the Owlman movie and was appalled at the idea that he had invited a bunch of people who make gratuitous lesbian art house movies to despoil the fair face of his beloved nature reserve.

- MAYBE he had just decided that the advent of a motley collection of monster hunters (and bloody hell, he hadn't even seen the pub-licity photographs we had taken which made us look like *Public Enemy* in their more paramilitary guise) was going to cause more disruption than was worth it to the smooth running of his organisa-tion.

- MAYBE the Belgian secret service had got to him as well. Not content with trying to assassinate me for defaming the good name of their foreign and colonial policy they had kidnapped Pat and replaced him with a clone who was planning to throw me to my certain doom in the remains of what had once been the largest lake in England.

Okay the last suggestion is purely facetious but all the other three and more had flashed through my paranoid mind.

However nothing of the sort had happened. Pat was merely preoccupied with some particularly intransigent paperwork which had to be com-pleted by close of play that day and so he had less entertaining things to do than to await the arrival of the first battalion of fortean fusiliers. He walked over to our table. Introductions were made all round and then, miraculously, Carl reappeared and I gave what I suppose was a formal briefing to all and sundry about what we were hoping to achieve and

191

how we were hoping to achieve it.

I also gave Past our insurance cover note and our signed disclaimer which absolved them in perpetuity of any responsibility for any mishaps, accidents or injuries to life and limb that might have resulted from any of our activities over the next few days.

Carl and Pat then shepherded us towards the place that was to be our home for the next four days. The New Raynes Observatory turned out to be the very same hide that Richard and I had first sat in six weeks before and photographed the lake. They showed us how to use the lights and power points and then left us to our own devices.

It took a surprisingly long time to unpack everything. Because we were camping in what was essentially a public area we had to make sure that we caused as little disruption either to the general public or to the staff and management of the reserve as possible. Therefore, although we un-earthed our technical equipment and the boat and addles and assembled them we decided that the food, camping gear and general domestic im-pedimenta should be left in the cars except during the hours when the public were no longer allowed on the reserve.

The electrical equipment, my computer, the cameras, walkie talkies and mobile phones were set out relatively neatly in the hide itself whereas the boat, now inflated and with the wooden deck and fishfinder rig installed was laid neatly together with the oars and the huge canoe paddle that Graham and I had spent half a day rushing around Exeter attempting to buy on the grassy slope that led down from the hide to the lake.

As we were starting to unpack the gear from the cars we had our first stroke of luck.

One of the main aims of the expedition was to get as many pieces of eye-witness testimony recorded for digital posterity as we could. The first witness that we managed to get hold of just wandered into the New Raines Observatory as we were unpacking. He was a late middle aged man called Hugh Page. I felt slightly sorry for him actually. The poor chap had gone to his favourite nature reserve for an afternoon's bird-watching and had found himself accosted by a bunch of bearded men in military uniform who had shoved a digital camera in his general direc-tion and asked (ever so politely) for an interview.

192

He had been one of many witnesses who had seen, what appeared to have been an attack on a swan:

FREEMAN: Can you tell us in your own words what happened?

PAGE: It was sometime around the end of February. I can't remember the exact date, but I was over in the swan lake hide. They feed the swans twice a day, and although they had finished feeding them I had called in on my way home. Most of the swans had wandered off, but one particular swan was still hanging around a bit. It looked as if it had been dragged sideways, and then it just managed to get itself free and scrabble up onto the bank.

FREEMAN: What species of swan was it?

PAGE: It was one of the Whooper swans.

FREEMAN: Did it show any signs of injury when it was up on the bank?

PAGE: It didn't, no. I think it must have freed itself fairly quickly. I knew before then of this "creature" lurking in the water, but I hadn't seen anything of it before

FREEMAN: Did it make any attempt to defend itself?

PAGE: The swan? No. Not really. Once it got out on the bank it knew that it was safe from attack and settled down

FREEMAN: What about when it was in the water?

PAGE: It was flapping its wings a lot. Something must have got hold of its leg and it struggled to get free and back to the bank

FREEMAN: Did you notice any disturbance in the water?

PAGE: No, neither before or afterwards

I was interested in the fact that Mr Page had heard about the "monster" previously and wondered how much his interpretation

193

events had been influenced by his previous preconceptions. This is something I have encountered over and over again in my short and ignoble career as a UFO investigator, but I decided to apply the techniques I had learned in that sphere of endeavour into this investigation. So as not to make my motivation obvious I butted in with an innocuous question before asking the one that I really wanted to know the answer to:

DOWNES: What sort of time of day was this?

PAGE: It was about this time.. round about four o'clock

DOWNES: So in the winter it would be getting dark?

PAGE: Yeah. I had been down the bottom end and was just on my way home...'

DOWNES: You said you had heard about the strange creature before?

PAGE: Only from being here, and hearing that there was something in the water that had been attacking swans on occasion...

We thanked Mr Page for his time and co-operation and he wandered off.

The time for talking was over, and it was time to investigate the lake itself.

The first thing that we had to do, now that we had untrammelled access to the Mere itself was to walk around the entire body of water and ascertain its physical characteristic. It has been said that wels catfish on occasion actually find themselves mud wallows in shallow water and even on mud banks at the sides of a lake, marsh or slow moving river where they lie in wait for hapless prey species to come within their fishy grasp.

Richard and Graham therefore carried out a fairly brisk circumlocution of the lake, in search of potential wels wallows but also in order to ascertain the basic geographical make up of the place. Much to our surprise it turned out that in the thirty years since the Wildfowl and Wetlands Trust had established the reserve no-one had actually carried out a thorough mapping exercise of the Mere, and that only one of the little islands that dotted the surface actually had a name.

Graham decided that this would have to change.

```
FROM GRAHAM'S LOG:

I spent about 1.5 hrs mapping the eastern half of the lake
(the half that's east of a line between the feed hopper and
Turner Island) The mapping was done in the traditional
"ancient mariners" fashion, ie looking at coastlines from a
distance and drawing impressions of them from various an-
gles - that's from a distance across the water and also
when on the island.  From any given viewing point one will
see some things better than others.  From many different
viewing points, a series of impressions are gained - the
raw data - which then needs to be collated, ie organised
into an overall map. This in some ways is like doing a jig-
saw - but with constantly dealing with anomalies. The map-
ping of the eastern shore and the above-water features of
the lake 250 ft or closer was concluded by about 4pm.
```

Richard and Graham returned from their circumlocution of the lake with
initially disappointing news. Not only had they found no signs of any
wels wallowing places, but in Richard's opinion the lake was too pol-
lutes by wildfowl droppings and too small to support a large aquatic
predator. If there was anything there during the previous winter, stated
Richard categorically, he was sure that it had swum off to pastures new.

Bernard Heuvelmans, the "father" of Cryptozoology, wrote in his classic
1953 book *Sur la piste de betes ignorees* (published in English as *'On
the Track of Unknown Animals'* a few years later), of "Cuvier's Rash
Dictum". The seminal French Zoologist Baron Georges Cuvier once
wrote that *"There is little or no hope of any more large land animals being discovered"*. This was in the nineteenth century and within the next
few years, the largest lizard in the world, the world's largest ape, and a
host of other sizeable creatures were discovered.

Ironically, the words *"This lake is too small to support a sizeable predator"* will forever be etched into the oral history of the CFZ as Freeman's
Rash Dictum, for not ten minutes later, as Richard was wandering about
on the eastern side of the lake he had two brief sightings of an enormous
fish!

The chase was now well and truly afoot.

195

Martin Mere mapping

When faced by a low-level wetlands area, any sand-banks or other bits that show above the overall water-line are likely, from ground level, to appear as a series of wide and thin pancake-shaped features at varying distances. The width of any given feature is "as seen" - however, the depth (extent) of it is severely foreshortened…GI

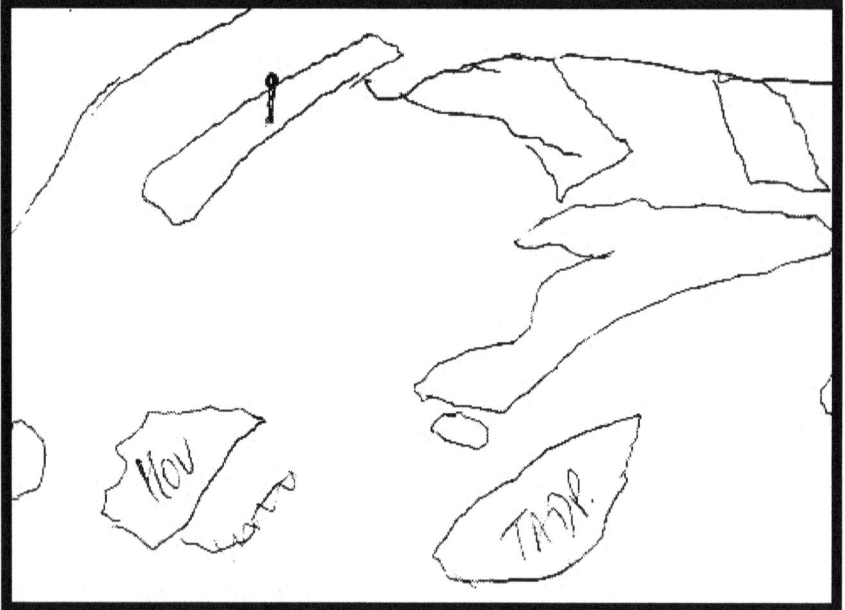

I'm 6 ft (1.8 m) tall, but this was insufficient elevation.

In the absence of an overview - and for mapping purposes an aerial photograph would have made all the difference - one either needs surveying equipment (poles painted alternating red and white, and a well-calibrated theodolite) or considerable imagination and a willingness to continually scout and re-scout the area to be surveyed. Or, I suppose, a *penchant* for out-of-body experiences.

Putting the latter option on 'hold', I pulled off my orange CFZ t-shirt and walked or waded from one "island" to another, the sun warmly shining down on me, hour after hour, as I created a succession of local maps. Try to gauge longer distances, and inaccuracy piles on inaccuracy - the pancake effect - so I soon realised I should stick to local mapping only, in any given spot. GI

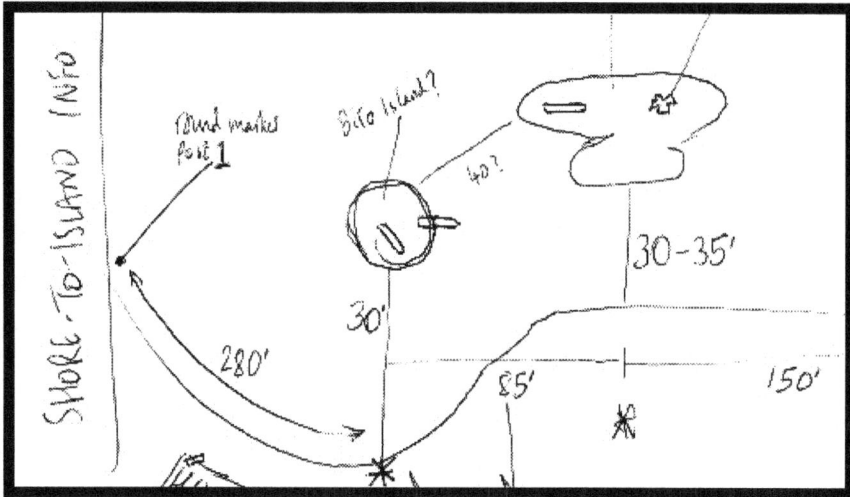

Stand on any particular sandbank or islet in Martin Mere and the distances to adjacent sandbanks can reasonably be estimated - typically, it was 50 to 100 ft (around 12 - 26 m) between one bit of dry land to another.

I later attempted to tie all these local maps into an overall map of Martin Mere, and this is where the problems really started. It was like doing a jigsaw puzzle, except that the extremities of each localised "jigsaw piece" had intrinsic inaccuracies that prevented them matching with the adjoining segment. The more distant the mapping, the more distorted and inaccurate the perceptions.

By now, I had considerable sympathy for the early mariner-cartographers, who were mapping continental coastlines, not just lakes. Maps drawn in Plato's time were distorted but still show the likes of Africa in recognisable form, and as I trampled around Martin Mere I developed more of a respect for these early efforts. GI

FROM RICHARD'S LOG:

I had noticed a deep water filled trench akin to a canal. Next day I was to follow its course. It was connected with the Mere via a small concrete pipe and was obviously a drainage device. When the Mere is full it would drain excess into the trench but at this time of year the trench was feeding the Mere.

The canal formed a rough square around the Mere and terminated in a pond. This pond was connected via more tubes to around ten further large ponds. Most of the connecting tubes were large enough to admit a fair sized catfish. Another of these ponds flowed in turn into the Mere. The ponds were surrounded by reed beds that stretched way into the horizon. The whole water course is connected meaning the creature could by anywhere on the trusts land.

That evening *The Ormskirk Advertiser* carried the following story:

SEARCH FOR GIANT PREDATOR WITH TASTE FOR SWANS:

Hunt goes High Tech for Mystery Monster

Experts from Exeter will spend four days using specialist equipment to discover whether Martin Mere really does hide a monster. Attracted by reports in The Advertiser earlier in the year about the possibilities of a wels catfish or a giant pike attacking birds the sleuths will be gazing over the waters of the wildfowl reserve's biggest lake. The Centre for Fortean Zoology will be unloading infra red cameras and sonar equipment tomorrow ready for its vigil. Martin Mere's centre manager Pat Wisniewski said the intention was to use fish finding radar-like equipment to try to trace large bodies of fish in the lake as well as to tow a baited line in the hope that anything rising to it could be photographed and identified just in case the mystery monster, which was seen dragging a thirteen kilo swan underwater is a wels catfish as has been suggested. The team will be using infra red sights to watch over the lake overnight for any nocturnal fish. At the moment there are no big stocks of visiting birds and most of the resident birds have finished breeding so there is nothing to be harmed out there. "It is an ideal time to do it" said Mr Wisniewski.

Angling experts were divided as to whether the waters of swan lake were deep enough to hide a giant catfish which can grow to 62lbs in Britain and reportedly 202 lb abroad, or whether it was a large pike trying to attack a swan.

After an exhausting evening during which, it has to be said, tempers got a little frayed we all disappeared to bed, me in the Jag, Graham in the Volvo, and the other two in makeshift accommodation on the floor of the hide.

Back in Exeter the following story appeared in the *Express and Echo*:

CITY'S STRANGE PHENOMENA TEAM ARE TO GO IN SEARCH OF THE 'MONSTER OF THE MERE'

BY PAUL MATTHEWS

12:00 - 24 July 2002
Jonathan Downes would not look out of place on an X-Files set. He is an expert on the supernatural and boasts vampires, giant snakes and Westcountry big cats among his specialist subjects. And tomorrow, Jonathan will be travelling to Lancashire to get to the bottom of more strange happenings in the depths of the mysterious lake known as Martin Mere.

There have been frequent sightings of a giant creature in the lake's waters, and many visitors have witnessed swans being attacked. Now Jonathan, 42, from Exwick, and other members of the Exeter-based Fortean Society, are determined to track down the demon of the deep.
In doing so they will be investigating treacherous and uncharted waters. The lake, near Ormskirk, is steeped in mythology. Among its compendium of spine-chilling stories are tales of dragons, unexplained ancient human mutilation and even a resident mermaid.

Martin Mere is also believed to be the resting place of the most famous of the knights of King Arthur's Round Table, Sir Lancelot. However, Jonathan and his team of Richard Freeman, 32, Graham Inglis, 45, and John Fuller, 40, are hoping for a more scientific explanation. He said: "Although sometimes I have to come to terms with the fact that I am dealing with things outside conventional science, this time I am not going up there expecting to find a giant monster. Instead, we are expecting to find a wells catfish, a species which should not be found in Britain."

It is thought that a wells catfish might have been introduced into the lake by Francis Buckland, of the Acclimatisation Society, during the 19th Century
.
The society, which was made up of a group of scientists, experimented in introducing exotic animals into Britain in

199

the 1850s. John added: "It is a lovely idea that the fish in Martin Mere is the same one that was introduced by Francis Buckland all those years ago. "This is certainly possible, as they grow very slowly, and some can live for hundreds of years. They can also reach a weight of about 300kg and get up to 16 feet long. If it is a wells catfish, it will almost certainly be a British record for the biggest fish in Britain."

Pat Wisniewski, centre manager at Martin Mere, believes that Jonathan may have found the answer to the mystery:

"We know that over the last century there have been illegal attempts to introduce the Wells catfish into British waters," he said "The giant catfish could conceivably survive in the murky, de-oxygenated water for years and grow to an extremely large size. This would seem to explain all the incidents which have occurred of wild swans being dragged under the water. On two occasions earlier this year something in the water has caused over 1,500 wild wintering swans to completely disappear from the lake. However, all the swans we have seen being attacked by the monster have survived."

The team of specialists will be using a range of hi-tech equipment to find the swan-snatcher. "We will be using electronic fish-finder sonar equipment, fluorescent rope and bamboo stakes to get a 3-D contour map of the lake," Jonathan said. "People may have seen us practising this on the river down by the Mill on the Exe.

"By doing this we will be hoping to find where the fish is lurking, and then we will buy mackerel heads and pull them along in a pillow behind our boat to bring it to the surface. We are also getting floating pond pellets for the other fish in the lake to feed on, to attract this big predator to come up to feed."

To clear up the mystery for good, John will be using military-style infra-red cameras and night-sights to get photographic evidence of the 'Monster of the Mere'
.
In recent years, the Fortean Society, which is 10 years-old, has taken its detective work to countries far and wide.

As Britain's expert Fortean Zoologists, they have travelled to Puerto Rico in Central America to investigate El Chupacabra, a vampire-like animal targeting goats. The team has also hunted dangerous giant snakes in the rainforests of North Thailand and researched Nevada's giant Thunder-

bird - the mythical bird icon of native Americans.

John is also sure of the explanation behind the mystery of the big cats on the Westcountry moors. "I have seen them and they are definitely panthers which have been released or have escaped from zoos," he said.

It would be churlish to enumerate the factual errors in the above story because the guy who wrote it was nearly as enthusiastic about the project as we had been. We took it as a good omen for the trip ahead and did our best to settle down to sleep.

Cartoon by Mark North featured in Animals & Men, issue 28

Friday 26th July

I am sure that my father who had been kind enough to give me the Jaguar as a present would have been appalled at the thought that I was going to spend four whole days living in it! I don't think that anyone had a particularly good night's sleep and relations between Graham and me were particularly strained this morning.

From Graham's Log:

Got woken up prematurely at 08.05 by people talking loudly - I should have driven the blasted Volvo off somewhere else. Felt pissed off. So Jon suggested I start the mapping, as this is a solitary activity and thus seemed likely to appeal - which it did. However, as an aspect of this activity involved being in radio contact, donning the "uniform" and taking the video camera, it counts as work and not chill time or relaxation.

However things were soon to take a remarkable turn for the better. After breakfast, as John and I tidied away the camping things and began to make a certain amount of order out of the chaos which reigned inside the New Raines Observatory, and whilst Graham wandered moodily around the other side of the lake continuing his mapping exercises (which by the way did eventually bear fruit in a remarkable fashion), Richard went off on a walk along the lakeside to the place where he had seen the fish the night before.

FROM RICHARD'S LOG:

At around nine in the evening on the first night I was
walking between a lakeside house and the swan hide when I
saw a portion of the back of a large fish that appeared to
be basking in the shallows. It was an oily black in colour
and bore a dorsal crest. The visible portion was around a
foot in length , six inches wide, and stood about four
inches out of the water. The creature dived swiftly,
alarmed by my presence. I flung bait into the water and the
creature's back surfaced once more briefly. It then disappeared. Gauging the size of the animal was hard as only a
small area of its back was visible but judging by the disturbance it caused it must have been substantial.

The following day at approximately 11 am I was in the same
area when I saw a large disturbance in the water. This time
the fish did not break the surface but it formed an "s"
shaped disturbance over three feet long. the commotion was
greater than that of the previous sighting. It was either
three or more fish swimming single file or one large elongate fish. Baiting failed to make the creature re-appear.

This was exciting. For the first time we had conclusive proof that there
was, indeed a large and hitherto unknown piscine predator in the lake.

Pat had furnished us with a complete species list of fish known from the
reserve. As can be seen, the fish seen by Richard (and indeed by other
witnesses over the years) does not correspond with *anything* presently
known from the area.

The fish that have been recorded from the reserve are:

PIKE *(Esox lucius)*
CARP *(Cyprinus carpo)*
CRUCIAN CARP *(Carassius carrassius)*
GOLDFISH *(C auratus)*
GUDGEON *(Gobio gobio)*
COMMON BREAM *(Abramis brama)*
BREAM x ROACH hybrid
RUDD *(Scardinus erythrophthalmus)*
TENCH *(T tinca)*
ROACH *(Rutilus rutilus)*
EEL *(A anguilla)*
THREE SPINED STICKLEBACK *(Gasterosteus aculeatus)*
NINE SPINED STICKLEBACK *(P pungitius)*
PERCH *(Perca fluviatilis)*

In addition the following species have been introduced over the years and are probably no longer present:

DACE *(L leuciscus)*
CHUB *(L cephalus)*
MINNOW *(P phoxinus)*
STONE LOACH *(Neomalichus barbatulus)*

When looking for the identity of the creature that Richard had seen we can discount a number of these fish immediately.

Pike are the largest predatory fish that are known to exist in Britain. However they have a distinctive shape and colouration and move in a diagnostic arrow-like fashion, utilising the strong muscles in the rear half of the body. Their characteristic jack-knife movements have earned them the alternative name of 'Jack' and is quite unlike the undulating motion of the fish seen by Richard. The pike is also heavily scaled, and has distinct green and brown stripes whereas the fish seen by Richard was scaleless and an oily black.

Carp would seem to be a more likely candidate for the mystery fish. However, although both mirror carp and the wild version are found in the Mere and can attain a considerable size the leathery species (which is scaleless) has vanished. Also, according to Pat at least, the carp that he has seen in the Mere are all brownish green or golden brown in colour, and so the colouration of the fish seen as well as the absence of scales would preclude any of the cypriniformes on the list.

The other species listed are all too small, and many of them also have distinctive colouration such as the green bands of the perch and the silver and red colouration of the roach). However, the fish species list does present us with some interesting anomalies. The most notable of these is the presence of gudgeon. These are bottom living fish of well oxygenated rivers, and how they can survive in the murky waters of Martin Mere is a mystery to me. Also interesting is the way that goldfish - a completely artificial morph of a fish found in the far east - have been found in the lake.

Something else particularly interesting is the occasional presence of Marsh Frogs - a European species usually only thought to exist in small numbers in the far south of England. They turn up in small numbers in

205

this part of Lancashire, and although it has been suggested that they are escapees from captivity, I think it far more likely that a great deal of revisionary work needs to be carried out on the herpetofauna of the region generally.

However, enough of my musings, we should return to the expedition.

Just as Richard was logging the events of the morning, the next stage of the weekend burst upon us. Up until now the whole affair had been between the Centre for Fortean Zoology, the WWT at Martin Mere and a bloody great fish. As far as the newspaper and radio were concerned, up until now we had been doing all the work. Just before lunchtime on Friday 26th July 2002, a jolly little bloke called Rob from the *Liverpool Post* turned up and the whole affair became a complete media feeding frenzy.

Regional news
Monster of mere may be catfish

By Scott Faulkner Daily Post Correspondent

THE "monster" of Martin Mere could be a giant catfish the length of a car, according to investigators

Specialists using hi-tech sonar equipment spent four days staking out the Burscough nature reserve after a fully-grown swan was seen being pulled under the water by a mystery creature. Several visitors witnessed the swan trying to flee the giant underwater predator in February. After an earlier incident, the 20-acre lake where swans gather was left deserted as they all refused to go on to the water.

Jonathan Downes is leading a team of experts from the Centre for Fortean Zoology in Exeter, who have investigated the Loch Ness Monster. He said: "We have been using sonar fish- finding equipment that bounces echoes off the water. "We are pretty sure we have found the area where it is and we believe it is up to eight feet in length.

"We have not got any pictures yet but are using nightsight camera equipment. If it is that size then it could be a record for a catfish in this country."

Centre manager Pat Wisniewski said: "Something is completely spooking the swans. On two occasions, January 17 and February 7, something in the water has caused the

1,500-plus wild wintering swans to completely disappear. Whatever it was out there last night must have been pretty big to pull a swan back into the water. Swans weigh up to 13 kilos. One theory is that the fish, which is not native to this country, found its way into the mere through its drainage system many years ago as a tiddler and remained there ever since having grown too large to escape.

Visitors to Martin Mere are asked to keep their eyes peeled

- *he British record rod caught catfish is 62lb.*

- *It can grow to a length of 16 feet but the UK record is nearly four feet.*

- *he world record is 102lb - dating back to 1890.*

- **The catfish is given its name because of its cat-like whiskers.*

I have no memory whatsoever of speaking to Scott Faulkener, and considering that this time it was me who was quoted as claiming that the fish was as *"big as a car"* and talking about sonar equipment in words that sound completely unlike anything I would actually say, it is quite possible that I never talked to him at all. I *do* however remember talking to a young lady with an incredibly sexy voice from the Press Association. I gave her a lengthy interview and she also talked in some depth to the Dark Earl of Gothdom.

Within hours the following report appeared on Ananova, and was posted to me by my mate Dave McMann:

"Hunters believe they have spotted a mysterious giant fish thought to be responsible for attacks on at least two fully-grown swans.

Sightings of the creature, dubbed the Monster of Martin Mere, were reported earlier this year by visitors to the bird reserve at Burscough, Lancashire.

Now a four-man team from the Exeter-based Centre for Fortean Zoology say they have seen a "very big fish" during a 24-hour watch at the Wildfowl and Wetlands Trust beauty spot.

Organisation members, who claim to have pursued vampires in Mexico, dragons in Thailand and skunk apes in Florida, have launched a four-day operation in an attempt to crack the mystery of the fish.

They are spending four days at Martin Mere using infra-red cameras, military-style night lights and "fish finder" sonar equipment in a bid to find out more about the mystery beast.

Director Jonathan Downes, who said he was somewhat sceptical upon first hearing reports of the fish, believes the monster may be a Wels catfish that could have been born during the reign of Queen Victoria .Native to mainland Europe and introduced into parts of the UK in the late
19th Century, Mr Downes said the Wels catfish is the largest freshwater fish in the world and can reach a length of 16 feet.

The "eight-foot fish" was first spotted on Thursday by team member and qualified cryptozoologist Richard Freeman who is hoping to capture the creature on film. He said: "I have seen something black and shiny snaking around in the water in almost the same place as the original sighting several months ago. It certainly looked like a Wels catfish.

"I can't say for sure that it was a Wels catfish. But if a pike had attacked the swans there would have been wounds. This thing seems to come up underneath and drag its prey down under the water."

Things were really beginning to come to a head. Within minutes of the story appearing on Ananova, BBC News 24 were on the telephone to me, and half an hour later this story appeared on their website.

"A giant fish which has attacked swans at a bird sanctuary has been spotted by wildlife experts. At least two swans have been hurt by the underwater creature nicknamed the Monster of Martin Mere which hides in a lake in West Lancashire.

Now a four-man team says it has located the attacker which they think could be a Wels Catfish from eastern Europe.

Although the biggest ever caught was 16-feet long, Jonathan Downes, one of the team from the Centre for Fortean Zoology in Exeter, thinks this one is a tiddler by comparison - perhaps seven feet long and weighing 24 stone.

The mere is home to scores of swans

He said: "If it is a Wels, it is almost certainly a British record." The team is spending four days at the Wildfowl Trust at Martin Mere using infra-red cameras, military-style night lights and sonar equipment to find out more about the mystery beast.

The giant fish was first spotted on Thursday by team member Richard Freeman who is hoping to capture the creature on film. He said: "I have seen something black and shiny snaking around in the water in almost the same place as the original sighting several months ago.

"It certainly looked like a Wels catfish.
'RUBBERY' BODY

"However we will be carrying out further investigations over the weekend in hope of obtaining photographic proof". Mr Freeman said the fish had no scales, had a *"rubbery" appearance, was oily-black in colour and moved quickly through the water. "I can't say for sure that it was a Wels catfish. But if a pike had attacked the swans there would have been wounds.*

"This thing seems to come up underneath and drag its prey down under the water." Reports of a larger-than-life creature living in the 17-acre lake were first voiced four years ago and the Martin Mere monster has since become a talking point among people living near the 380-acre reserve which regularly attracts Hooper[sic] and Bewick swans. "

This was my first introduction to the real operations of the high speed modern mass media. Although I have been online since the end of 1997, and have been using emails and the worldwide web pretty well on a daily basis since then, and although my own working practises had been irrevocably changed by the advent of the new information technology, I don't think that until today I had grasped quite what an enormous effect the information superhighway had actually had upon the business of newsgathering.

My first forays into the world of the media were in the late 1970s working as a publicist with various punk bands. It was then (with some invaluable help from my father who would have been appalled if he had known where his help on matters of press releases was going to lead), that I cut my eye teeth in matters of media manipulation. It wasn't however until over a decade later when I found myself in the unenviable position of working for one of my lifelong idols - Steve Harley - that I really learned how to deal with newspapers. Harley had been a journalist himself before becoming the most literate and articulate pop singer of his generation. Sadly, by the time I worked with him, he was no longer selling records in the numbers that he had been between 1974-6, but he taught me most of what I know about the press. Indeed, if it had not been for him, the CFZ as we known it (a bastard mixture of scientific credibility with rock and roll attitude) would never have existed. Unfortunately our relationship did not end too well and I never got the chance to thank him for everything he had done for me. Not until now.

I'm sure Steve (who taught me the fine art of conducting half a dozen interviews at once), would have been so proud of his protégé as I went through interview after interview over the next three days. By Wednesday either Richard or I had done something approaching sixty of them - and I hope that each time, as Steve Harley had taught me all those years ago - I managed to make the person asking the questions feel that those questions (and indeed THAT newspaper/radio station/or whatever) was of paramount importance to me.

The business of giving interviews had never changed, but the trappings of the job, and indeed the methods by which the information was assimilated into the public domain had changed beyond all recognition. I had been brought up in a methodology which started off with manual typewriters and letraset, and I was now working in a medium whereby even the most insignificant word I said could be instantly transmitted across half the world. It sometimes became too much for one little guy like me to deal with.

While I fielded interview after interview, Graham continued with the mapping and Richard and John patrolled the lakeside in the hope that we would manage to clock up another sighting of the fish.

The downside of the day was that I had started smoking again!

At Christmas I had decided that if I was actually going to have any chance of living a few years longer I would have to do something about my health. I quit smoking a few days before New Year's Eve, and after taking advice from my medical professionals I started a stringent health regime which included a severe reducing diet, and the complete absence of caffeine and refined sugar from my diet. I even quit alcohol, which is something that would, I am sure, have amazed many of the people who have known me over the years.

One of the strange things about being a fat man is that in the end your body shape begins to take over your whole personality. I have been grossly overweight for at least twenty years and have been known to all and sundry as "big John" for most of that time. Now, for the first time in my life I was making a real, concerted effort to lose weight, and in order to do so it involved making radical changes to not only my lifestyle but my self image as well.

I began to get used to my new lifestyle after a few weeks and I am proud to say that whilst I was at home I hardly strayed from the paths of righteousness. When I got to the lake the results on my efforts began to reveal themselves to me. Whereas during our first visit to Lancashire two months before I had been so ill that I could hardly walk. Now, I had lost a considerable amount of weight and my medication was having considerable effects upon my heart problems. This was probably not the right time to embark on an expedition, because almost without thinking I slipped into my normal expedition diet of stodge, smokes and spirits.

It is a diet upon which I have always thrived during my periods either in the field or on the road with various rock and roll bands during my misspent youth. However, it was a diet which I can only keep up for a short length of time.

As I sit up in bed writing this, a week after our return, my body is feeling the ill effects of returning to a responsible diet. For some reason what live on perfectly whilst I am out and about is poison to me when I am at home. So whereas a return to decaffeinated tea, artificial sweeteners, fresh fruit and yoghurt, and an absence of booze and tobacco is a bit of a jolt to the system, it will be a good thing in the long run (hopefully it will ensure that there will *be* a long run), I know full well that when I am next in the field, out will come the bourbon and the chocolate and the Benson & Hedges, and my world will continue as it always has.

```
FROM GRAHAM'S LOG:

Another photographer got pix of us all - with me & RF in
the dingy in the background. The Volvo's front offside tyre
is flat. Jon reckons repetition of types of mission should
reduce preparation frazzle as once we've got the gear, then
it's available for reuse. I videoed Bernie Starkie (RF off-
camera interview) at 5pm.  Doing this log and Jon's just
walked over to join me, but has now received another phone
call and is talking
```

That night I drove into Southport and bought Chinese food and whisky. After dinner, replete with Duck and orange (from the same vendor we had visited on our first night in Southport back in May), and with a healthy amount of Bourbon swilling around my veins, I sat back and looked at the lake. Making a silent promise to myself that I would give up smoking again once I returned to some semblance of normality, I lit a cigarette, laid back on the grass and looked up at the night sky. The ducks were making chortling noises to each other in the middle distance and the stars were twinkling. For once I wasn't even fighting with Graham.

Surely life doesn't get much better than this.

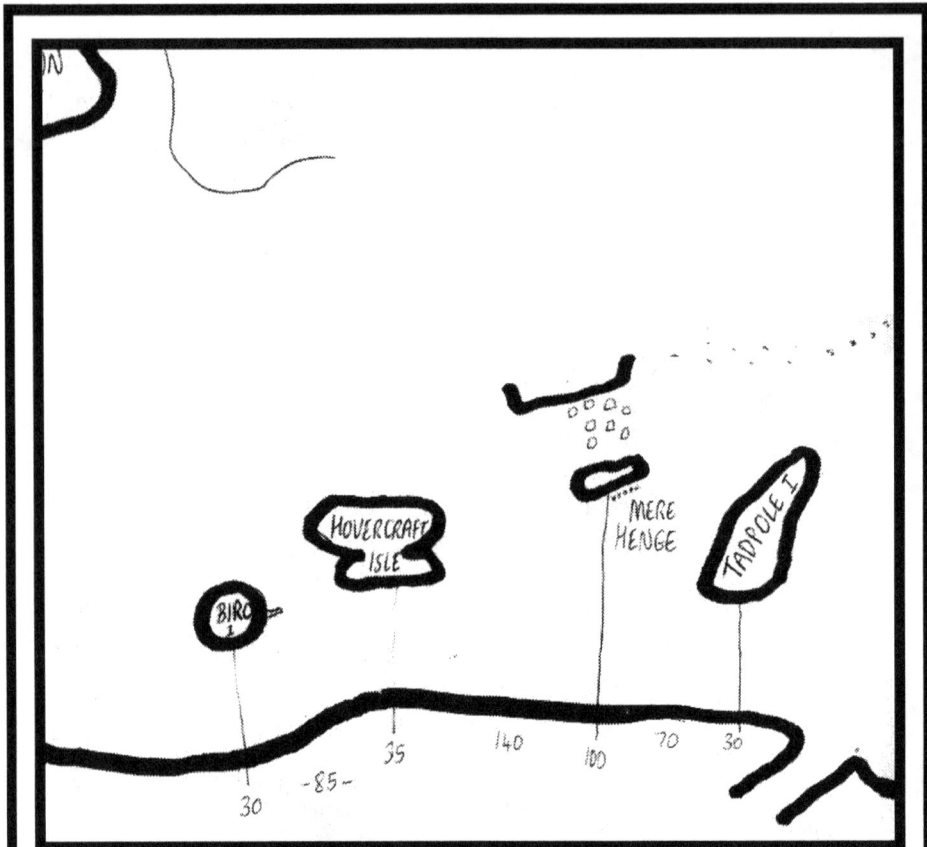

In my opinion, an inaccurate map can be worse than no map at all, as it's only human nature these days to initially trust a map when presented with it. So.. no cop-outs were acceptable, as I felt that Jon and Richard wouldn't be satisfied with just a shoreline diagram and the word "unknown" inscribed acoss the watery bits - and substituting the phrase "here be monsters" might have envoked traditional cryptozoological *zeitgeist* but wouldn't have advanced the CFZ cause right here and now. Anyway, we already knew there was a 'monster' there - Rich had seen it a couple of days ago.

I sat by the lake, lit a cigarette, gazed at the murky and enigmatic waters, and mused on how accurate the map need be. My mindset was that of someone at a logistical disadvantage trying rapidly to get onto more even terms with a target that has had goodness knows how long to figure out its options. Since we were tracking a creature that may well have existed in this lake (or Mere) for decades, it seemed to me a reasonable assumption that it knew its own "back yard" pretty well: a boat approaches, say, and it knows where the deeper escape channels are. In short, it's at home.

Although we were not aiming to catch the thing, still less kill it, I still felt that the required mindset was that of a hunter, at least in the initial tracking sense. We weren't going to shoot the target, but tracking is tracking, no matter what your intentions are when you find your prey. It logically follows that the more you learn about the local terrain, the better your chances of finding it. In my opinion, having an accurate map of the islets and water channels was a crucial element in the tracking process. Effectively, the map would bring us up to speed with the options long known about by the target. So, it behoved me to make the map as accurate as reasonably possible.

All well and good, but this musing brought me no closer to actually creating the desired map.

One important factor with this particular lake's profile is seasonal fluctuation - the one pre-existing map I had access to was a vague tourist one, drawn up at near high-water mark - a rainy season map - and it took me several hours before I could fully correlate the map with the current summertime vista. The eastern shore is packed with landmarks all year round - several observation hides, a binoculars shop, and a 30 ft high food hopper - but the rest of the shoreline is devoid of any absolute reference points.

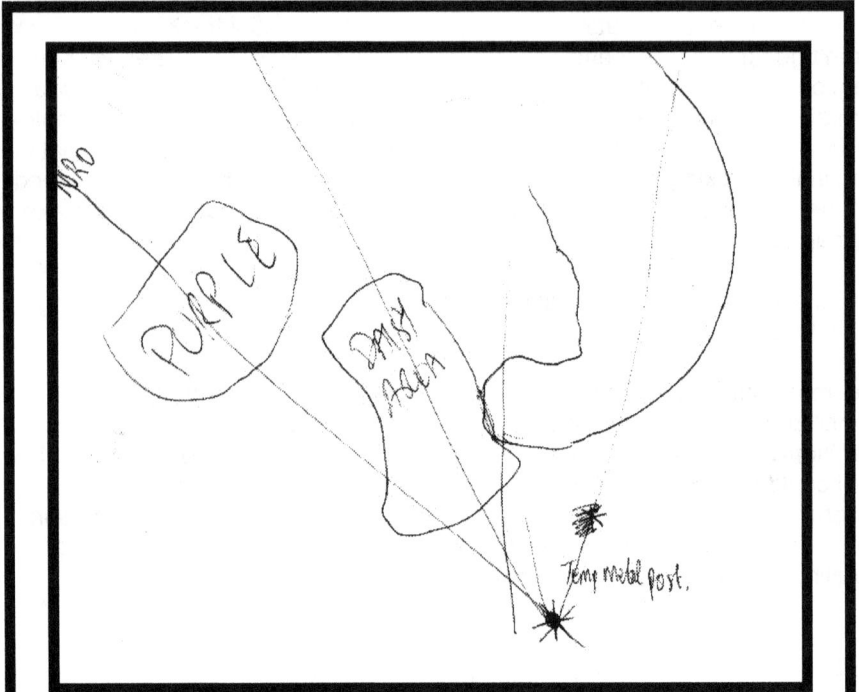

Time for some baseline data.

Standing on the shoreline, I decided to pace out the overall dimensions of the lake and then complete the elusive and rather tiresome island details afterwards. I'm lucky in that my stride is almost exactly one yard - 0.9 m - and in this way I determined that the lake was - if I pretended for the time being that it was a perfect rectangle - around 700 x 400 ft. Any handle on a problem is better than no handle at all, so I opened a can of beer and roughed out an initial impression of the area - a "working diagram", so to speak.

That competed, I then reflected on the way ahead.

One problem encountered was bizarrely prosaic: in the few days we were there, a National Trust pumping problem less than a mile away caused the water level to sink by 6" (15cm) - a fall of around 15% - and that made a considerable difference to the overall topography. Some islands could now be reached wearing ordinary shoes, whereas previously one would tread cautiously in wellies, not knowing how much silt lay under the sparkling water. Islands previously isolated were now connected by spits of gravel.

Saturday 27th July

Looking back on the events of the few weeks is quite a task. In suburban Exwick, in my darkened room only lit by the screen of my laptop and the flickering of the silent TV in the corner, with Marianne Faithful on the CD player, the lake seems a long way away. Lisa, my daughter had just come in to kiss me goodnight and Helios 7, the cat, is asleep on the end of the bed. Trying to make sense of our adventure from the point of view of the middle aged man lying back on his elbow, typing with one hand as he waits for his cyber lover to come on line five time zones away is a difficult task. So I put on the unedited dub tape of all the video that Graham shot during the week.

Watching it, I suddenly realised how much the pace had changed by mid afternoon on Friday. Whereas the first two days had been ones of acclimatisation, by this time we were not only ensconced firmly within our camp, we had also begun to make our mark quite incisively upon the landscape of Martin Mere.

FROM GRAHAM'S LOG:

Yesterday I planted the canoe paddle on Ali Island as a cartography reference point. Today I will use it to help me assess the overall shape of the mere. Distances between points were assessed by means of pacing them out. Richard reckoned the tree is 16-18', not 24'. I'll look into this matter shortly and then seek to make the master chart de-

ferred from yesterday. (I had some unexpected peace and
quiet when the others went shopping in Burscough and util-
ised much of it in having a chill.

Later I donned the waders and took a rope over to Biro
Isle. I was squelching through roughly 18'' of silt and
18'' of water, probing the forward terrain with a walking
stick. (Amusingly, I'd earlier gone off on my own to explore the feasibility of reaching Biro Isle on foot and
didn't even bother with a radio or lifejacket.) Stakes were
hammered into the ground either side, and I then stretched
the rope between them.

I then took the fish detector transducer out to the middle
of the channel and suspended it from the middle of the
rope, leading to the fish-finder on the bank. After power-
ing it up and doing a few tests, I then left RF and JF to
keep watch.

Graham's mapping operations were continuing apace. I started this narrative by making oblique references to Arthur Ransome, but this adventure was by far the most Swallows and Amazonseque adventures of the
CFZ's career to date.

His novel *Secret Water* was, according to Bill Wright's exhaustive
analysis of the Swallows and Amazons books, set in the summer of
1932. In this, the eighth book of this classic series, the five Walker children (a.k.a. "The Swallows") are left on a "desert island" by their parents
with provisions for a long stay and a blank map to fill in. Like all Ransome's books, this is at once a real adventure and a lesson in the practicalities of exploring - in this case, of surveying the inlets, coves, mudflats, and estuaries of "Walker Island." Although the book is set in the
mudflats of coastal Essex, the parallels with our expedition to Lancashire
are irresistible.

I was surprised to find that I was not the only member of the expedition
to have read *Secret Water*. Neither Richard or Graham had read it, but
John - to my delight - was an avid fan of all the Ransome novels. John
contradicts every preconception of the cultural tastes of a long term unemployed black man in contemporary Britain. Not for him chilling with
his posse, when there are the delights of P.G.Wodehouse, Arthur Ransome or Bob Dylan to grok in their multifarious fullness. As least in
John I had one cultural ally to share the adventure with.

Graham spent much of the day continuing with his mapping exercises. Even relatively early on the various features of the lake were given names. There was an Ally Island (named after a girl of whom Richard is somewhat enamoured), and a Rebecca's Channel (the place where the "monster" had actually been seen) named after my inamorata in Maryland. However, some of the islands were named by Graham in his own inimitable fashion. Tadpole Island (named because of its shape rather than because it was a breeding ground for *Rana temporaria*), Hovercraft Island (because Graham fancied it was the shape of a hovercraft) and Biro Island ('cos it was small and round like the view of the top of a biro), were all the results of Graham's own peculiar type of nomenclature.

Despite the fact that we have had our differences over the years (and particularly on this expedition) Graham and I are close friends, but I will be the first to echo the words of my elderly father. Inglis is a nice chap....but he is a most peculiar fellow. For example, when we were in Mexico (on our adventure recounted in *Only fools and Goatsuckers*) we were visiting the house of renowned Mexican UFOlogist Jaime Maussan. He lived in an extraordinary house high in the hills above Mexico City. His house was a collection of handcrafted log cabins linked with glass topped tunnels. The cabins were covered with exquisite hand carved depictions of Mayan Gods and strange animals and they were highly polished to produce a patina I have never seen in exterior woodwork ever. All around the house were delightful gardens and everywhere you looked hummingbirds were flying.

I stood entranced and tried to grok as much of this fantastic building as I could in order to make a permanent impression on my mind. I looked around to find Graham. He was nowhere to be seen. Upon investigation he was on the far side of the building totally ignoring the view, and the carvings and the hummingbirds.

He was taking a close up picture of the air conditioning apparatus!

THIS is truly the mindset of the only person in the world who could have named "Hovercraft Island".

We established a mini base on the shore near to where Richard had seen the big fish on the two previous days. This consisted of two plastic chairs, a pile of equipment, and suspended on a rope between the shore and Biro Island, the transducer of the sonar fishfinder right in the middle of Rebecca's Channel. We christened this outpost FishBase1 and for most of Saturday daytime, while I fielded more and more media requests from the base camp at the New Raines Observatory, and Graham continued mapping the rest of the lake, Richard and John manned FishBase1. Intermittently I would radio Graham and Richard to return to base and interview another eyewitness or talk to a reporter, and they would leave John manning the sonar equipment with quiet diligence.

It was during one of these periods when John was manning FishBase1 alone that we had out last major breakthrough.

FROM JOHN'S LOG:

As Graham and Richard were mapping Ali Island, they left me alone at FishBASE 1. I moved the plastic chair right down to the little gravel beach and sat down next to the fishfinder. As the sun was hitting the water and warming up rapidly the fishfinder began to beep repeatedly. I picked up some extremely large fish images on the viewfinding monitor and although some may have been shoals others appeared to be a large, single fish. There seemed to be more and more bubbles rising to the surface as if something was swirling around in the ooze at the bottom of the channel. The depth readings were going crazy as the beeping was getting more intense and the single fish image kept reappearing on the monitor. There were large ripples and disturbances on the top of the water as if something extremely large was swimming around at the bottom of the channel. Approx 11.50 a.m

About ten minutes later I saw two enormous splashes as a large animal, presumably a fish, leapt out of the deep water between Tadpole Island and Hovercraft Island, falling back into the water. It may have been my imagination (as I was sitting at the laptop in base camp at the time wearing my reading glasses, this observation doesn't count for much), but I am pretty convinced that whatever the large fish was, when it submerged it took a small teal with it!

Saturday was a much busier day as far as punters were concerned than the previous day had been. This is hardly surprising because most people work for a living and only have the weekends free to wander about their favourite nature reserves. Once again I have to apologise to the general public and thank them for their forbearance when they go our for a day's birdwatching and end up having a fat guy with long hair and a walking stick or a bunch of bearded blokes in battledress point cameras at them and ask for interviews.

I had been woken at an appalling hour that morning by a polite little Arab man that I have known vaguely for years. He runs a British Government sanctioned pirate radio station called *The Voice of Free Iraq*, and he broadcasts news from the free West to the beleaguered folk suffering under the yoke of Saddam Hussein and the Ba'ath Party. This guy (whose name I can never remember) is somewhat of a fortean at heart, and never misses a chance tom interview one or other of us for his station. I am always happy to help spread the word of truth and justice, but when it comes to being rung up on my mobile at 6.30 in the morning and asked damn fool questions about catfish, when you are lying naked in the front seat of your father's old Jaguar wrapped in a quilt, ones tolerance for the forces of freedom in the Middle east starts to rapidly vanish.

After I had successfully concluded this interview I decided that the call of nature was too strong for me and I went off for a pee before resuming my slumbers. Pat had been kind enough to allow us access to a small concrete lavatory building which presumably was usually locked for the night. As I ambled back to the care, vaguely wondering to myself whether I was going to have a cigarette and a cup of coffee before returning to my bed in the car I saw an extraordinary sight before me.

A young stoat – peculiarly dark in colour – was cavorting across the path in front of me. I have read accounts of people seeing stoats "dancing" before but this was the first time that I had seen it for myself. The tiny mustelid was leaping into the air with the sort of *joi de vivre* that one expects from small children or rock musicians. As it jumped, it twisted its lithe little body into the most extraordinary contortions. A brimstone butterfly fluttered across the path and the miniature carnivore leapt into the air higher than ever to catch it. Missing it completely, the stoat fell to the ground in a series of somersaults. Then it saw me. It stiffened and then hurried off as if ashamed that it had allowed me – a mere human – to witness its delightful display of acrobatics.

Then I went back to the car and soon I went back to sleep, to be awoken an hour or so later by the Radio 4 Today programme who conducted a lengthy interview with me and then proceeded to broadcast bits of it on their news bulletins every hour on the hour throughout the day. This interview also found its way onto the Classic FM News (where my father listened to it bemusedly as he drank his breakfast tea back in Devon), and several other radio news programmes besides. The upshot of this was that a steady stream of people, somewhere between a trickle and a throng descended on the New Raines Observatory wanting to talk to us, and a surprising number of them had information that they wanted to share with us.

One of the first people that we wanted to interview was Bernie Starkie whom we had met upon our first visit and who was highly sceptical about the stories about swans being attacked by a mystery predator.

FREEMAN: Ok Bernie, could you tell me about your theory about the swan?

STARKIE: Last winter there was a swan around for some time that had something which I sort of related to pictures of cattle with BSE. As soon as it got in the water it would flip over so its legs were almost up in the air, and it would try to get out and flap frantically to try and get upright. But as soon as it got out of the water it would appear perfectly normal. A bird trying to get out of the water to the safety of the bank wen its body is upside down struggles a bit.

The other birds didn't like this - whether they were worried that it might attract predators I don't know - but when it struggled they would clear right away. When it was trying to get out of the water there was a lot of splashing and I could easily see how this could have been interpreted as something trying to drag the bird underwater.

FREEMAN: Do you think that this could have been the reason that all the birds were "spooked" and left the water?

STARKIE: I have seen that happen, but by the same reckoning other people have seen the same thing happen for no obvious reason.

FREEMAN: Do you have any explanation for the big fish which Pat and other people have seen on the Mere?

STARKIE: No

FREEMAN: Is it your feeling that people are putting two and two together - linking two completely unrelated phenomena?

STARKIE: That is exactly what I think. I have been here four or five days a week, every week for three years now. and I have never seen anything inexplicable, but having said that I have seen extremely large bow waves moving through the water and until now I had put them down to the presence of large - but not *that* large - fish...That being said there are things out there all of the time that we

220

don't see most of the time, and there's absolutely no reason why *something* could not be out there.

However, many of the other people we interviewed took a completely opposing view. We spoke to a young man called Lee Bailey who had been a regular visitor to Martin Mere for many years. He was an articulate bloke with a cropped haircut and only one leg. He spoke at some length about why he was convinced that the swans involved in these incidents were neither ill or suffering from lead poisoning:

FREEMAN: Can you just tell us in your own words, what you saw, what time of year, and whereabouts?

BAILEY: I believe that it was about March this year, something like that, and it was in the Swan Leat hide which runs along one side of the Mere. There were a lot of swans on the Mere at the time. There were five or six other people in the hide with me at the time as well. I was being told about the original attack on the bird when the swans were being dragged under. One feller pointed out that there was a swan in distress. Something had grabbed its rear end and it was at an angle. It wasn't under its own steam, but was being dragged by something who was either pushing or pulling at it under the water. Eventually it managed to break free and struggle to the bank looking a lot worse for wear. Missing a few feathers and that sort of thing. I didn't see the animal or whatever. Whatever it was it was under the water.

FREEMAN: What happened then?

BAILEY: It made its way to the bank and sat there exhausted. Then it preened itself and so on and then got up and walked away quite happily about ten minutes later. But everything else had scattered. Pochards, tufted ducks. Everything that was on the Mere at the time just made its way onto one of the islands or onto the bank.

After the interview was over, Lee hung around and chatted to us as we prepared to film the next interview. As we were chatting he brought up the subject of swans suffering with lead poisoning and was so interesting on the matter that I asked him whether he would submit to a second interview. He grinned and agreed so Graham set up the cameras again...

DOWNES: You have seen birds which are suffering from lead poisoning?

BAILEY: There have been quite a number if these swans here.. again, usually by the Swan Leat. They come on the bank its like they are punch drunk. They can't stand up. They lean forward to preen or whatever and they just have no balance whatsoever and they just fall over and droop. They are very weak. Whatever had that swan that I saw, it was fighting back. If it had been poisoned it just wouldn't have had the energy to fight back the way it did.

DOWNES: There have been some reports that there was at least one swan suffering from some form of degenerative brain disease which meant that it wasn't able to stand upright.

BAILEY: No I've not heard of anything like that...

Lee was accompanied by an older man called Ken. As Ken's surname was also Bailey I assume that they were related but I never got around to asking. However, Ken also had a story to recount:

DOWNES: so what happened, Ken?

BAILEY: It was about four years ago, we were in the Raines Hide here. It was in the winter so the water was right up to the bank. As the geese etc were feeding at the end all of a sudden this big dark coloured fin appeared for a few seconds.

DOWNES: How tall was it?

BAILEY: It was about five or six inches in depth and perhaps nine inches long. But you've got to remember this was four or five years ago and so it could have been bigger than what I am thinking of. It was only a matter of fifteen or thirty seconds or whatever. It just appeared, seemed to roll over, and then disappear back into the water. The birds that were there scattered but didn't actually leave the Mere completely.

DOWNES: Had you heard the stories about something in the lake? Either before or since?

BAILEY: I hadn't heard anything before. All I have heard are the stories that have been going about since, but I had forgotten about the incident until today. I never connected it with the "shark" or whatever it was that people were originally saying that it might be.

All the time this was going on Pat Wisniewski was bustling around with a bemused look on his face. What on earth was happening to his beloved nature reserve? At one point representatives from a well known Satellite TV News team had got quite nasty when both Pat and I had refused to let them land a helicopter onto one of the islands in the lake. This was really getting out of hand. It seemed that every quarter of an hour Pat would wander up to me with an apologetic grin and another request for an interview written on a post it note in his hand.

I did interviews with Reuters, the *Sunday Times* and the Press Association in quick succession. By tea time I had done over thirty five interviews. In the early evening, once the main work of the day was finished I had a telephone call from our friend Nichola (The CFZ Fund-raiser) back in Exeter. It appeared that our exploits had been included in the International News section of the ITV Teletext service. As we had no access to a television set, she dictated the story down the telephone to me and I relayed it to the rest of the team:

"Monster Hunters believe they have spotted a mysterious giant fish at Martin Mere, west Lancashire, thought to be responsible for attacks on swans"

This was given equal validity in the news stakes alongside the doings of Gerry Adams, Dubya, and Yasser Arafat. Surely the world had finally gone completely mad!

We then took a couple of hours off with the idea that we would then resume activities about an hour before dusk. Graham wandered off to do his own thing whilst the rest of us sat around on the grass slope which led from the front of the New Raines Observatory town to the shore of the lake. For the next two hours we drank, smoked and looked at the dusk until the allotted time of seven thirty when we radioed Graham to tell him that it was time to resume fish hunting activities.

For some reason there are a number of spots along the shoreline of the lake which are radio black-spots and by sod's law Graham was sitting in one of them. It took nearly twenty minutes to locate and retrieve him, by which time it was beginning to get dark, However, as Graham launched the boat and made his way as slowly and gently as possibly from the shoreline by the observatory, round the far sides of Hovercraft and Tadpole islands and finally to the far side of Biro island, Richard, John and I (clutching cameras, night sights and video gear) walked through the dimpsey towards FishBase1.

Everything that we had researched about wels had convinced us that not only did they not budge particularly far from their home base but that they were crepuscular feeders, and that as we had established (to our satisfaction at least) that "Marty" had taken up residence in Rebecca's Channel, then we hoped that some intensive baiting would manage to coax him into view.

"Marty"? Yes, by now the Monster of the Mere had been given a name. Although Richard and I both abhor the trivialisation of cryptids through popular nomenclature we knew that someone would come up with a stupid nickname for the catfish sooner or later (probably sooner) and that it might as well be us.

The creature of Loch Ness had been christened "Nessie". The creature of

Loch Morar "Morag". The Chesapeake Bay sea serpent had been called "Chessie", and the monster photographed in Lake Champlain by Sandra Mansi in 1977 had been christened "Champ".

What could we call the creature that we were by now certain lived in the depths of Martin Mere? Because of the location's proximity to Liverpool, we were originally thinking of calling it "Scally", but as soon as we came up with "Marty", the name stuck, and by Monday lunchtime it was splashed across most of the national newspapers. At least, we thought, it could have been worse. We were thinking of whatever sick person it was who had christened the monster of Bungalow Beach in the Gambia "Gambo", and had coined the name "Trunko" for the strange trunked sea creature reported off the coast of Natal a hundred years or so ago.

It took quite a considerable time to get Graham positioned on Biro island where he sat perched on an old log, looking for all the world like a bad tempered and rather dishevelled cormorant. In the meantime Richard had gone for a short wander and had seen the large fin, or what appeared from its colour, to be an enormous carp break the water.

It soon became obvious that our quarry was not the only crepuscular feeder in the lake. As the cool night air began to flood over us like a fluid blanket, the tiny insects that all day had been riding the thermal currents high above the lake suddenly began to plummet to the surface of the lake which soon became covered with little ripples as the smaller piscine predators of Martin Mere gathered for their supper.

We hoped that this feeding frenzy of tiny fish would attract the slightly bigger fish and so we started to bait the surface of water with handfuls of floating fish food pellets all of which even the littlest fish of the lake completely ignored. Richard returned to us clutching a very dead lapwing that he intended to use as his *piece de resistance* in the baiting arena.

Because of our singular lack of success over the previous few days whenever we had used the floating fish pellets that Pat and his merry men had kindly provided us with we decided to try using a pungent 'chum' mix of our own invention. 'Chum' is a fisherman's term for a wide range of nauseating bait substances used to leave trails across the ocean by people seeking to attract large carnivorous fish, particularly sharks.

This usually consists of dead fish and portions of the same which are far beyond their human sell by date and are only of use in attracting the most voracious predators of the ocean.

We hoped that we could utilise much the same strategy during our search for "Marty". Earlier in the day Richard and John had prepared a nauseating mixture of mashed up corned beef, sardines and milk and left it outside to fester. By the time we were ready to use it in the middle of the evening it was stinking to high heaven, and we hoped would be irresistible to any self-respecting catfish.

John was positioned by the fishfinder monitor on the shoreline and I was behind the video camera as Richard threw the first few ladles of the disgusting mess into the water a few feet from the transducer. It was obvious that we had got the recipe right because quite a few fish were picked up on the fishfinder sonar as they sped their way towards this new and delicious addition to their menu. However, although we ladled in a whole bowlful of this pungent and revolting mixture into the long suffering lake, at various locations within about ten feet of the spot where the transducer entered the water it was to no avail. We saw several sizeable bream and other fish but nothing even approaching the monster.

We decided then that we would have to go to plan B.

Graham relaunched the dinghy and lying on his tummy paddled slowly up and down the bottom entrance to Rebecca's Channel with the huge canoe paddle swirling up the muddy water beneath him.

Richard continued to use the floating bait liberally on the water surface between the boat and the transducer, in the vain hope that a mixture of ground baiting and a desire to avoid Graham's disturbances on the lake bed would force the big fish into making his existence obvious. But it was completely unsuccessful and we eventually gave up.

However, just as John and Richard helped Graham bring the dinghy (by this time rather battered and the worse for wear) ashore and as I packed up the camera equipment and tripod we could see an enormous wake as if something very large was moving just under the water in the totally opposite direction to where we had been ground baiting. Whether it was "Marty", a large carp or just a figment of our imagination we didn't know but it was by now too dark to find out, so we shrugged our shoul-

225

ders and went back to camp where we made our plans for the next day.

FROM GRAHAM'S VIDEO LOG:

Jon did a really good job of fielding the media and the
public while we got on with the real work such as mapping.
Once we have mapped the channels we know the beast's options. We are not trying to kill the thing but we are hunting it so we have to have a hunter's mindset. It's been in
this lake perhaps fifteen years, perhaps more, and it knows
the area pretty damn well. What we have to do is to try and
approach the knowledge of what this beast's options are. If
we go charging over the lake at random then it will run
rings round us - probably literally - and so our knowledge
of the lake is very important. I've drawn up our chart of
the lake now, and tomorrow I want to do some waterborne
mapping and do my best to find out the depth of the channels and so on. If you are going to explore territory it's
handy to have a chart, and by tomorrow we will.

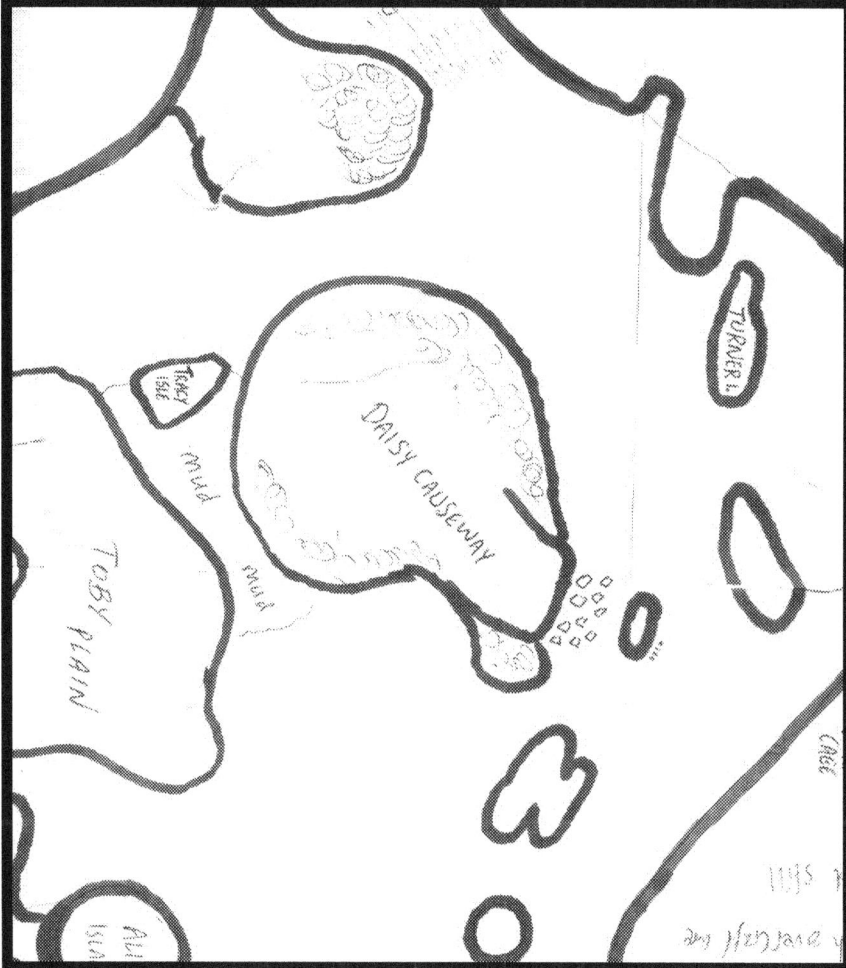

Shrugging off the water level issue, I decided to create a "snapshot" map of all the islands at this specific point in time - and if it poured with rain afterwards, or they mended the pump, then so be it.

Determined to create absolute frames of reference amidst all these foreshort-ened "pancakes", I walked off with the canoe paddle, which had bright orange blades, and strode around to the far end of the lake, and onto a gravel bank that Richard earlier had named "Ali's Island". I stood the paddle shaft upright and built a cairn of stones around the base to hold it steady, and then spent over an hour wandering around various parts of the lake, taking sightings of the distant paddle blade and having an occasional can of beer as I spread another sheet of paper out on the ground and wrestled with the changes that this reference marker forced me to make to my cartographical jigsaw.

Sunday 28th July

S unday was an incredibly frustrating day for the most part. It did, however have some brilliant moments, although on the whole it goes down in my memory as the day tat we lost control of the investigation. The day started well enough with the arrival of Sarah Burdon, a very cute young lady from Champion newspapers, who had (incidentally), been the person responsible for the amusing cartoon of Richard and me being attacked by a giant catfish which had accompanied Geoff Wright's article about us back in May.

She turned out to be as intelligent as she was attractive, and furthermore one of the few female journalists who have ever come to interview us who was not in the little bit "girly". She soon became co-opted onto the team and we gave her an honorary CFZ T Shirt and she joined in our activities for most of the rest of the day, often wearing John's battledress jacket. Like Lazarus Long is reported to have said *"what a wonderful world it is that has girls in it"*.

The first order of business of the day was to sort out the last few interviews that needed to be done in order to build up a complete picture of the events at the lake which had prompted our arrival and involvement in the first place.

Feeling horribly churlish for interrupting Pat during one of his few days off, we grabbed him as he was out walking his dog (who looked chillingly like my old hound Toby who died during the summer of 2000 and

229

whom I still miss dearly) and shoved a camera in his general direction:

FREEMAN: If you could just tell us about your original sighting...

WISNIEWSKI: Because I live on site, I obviously spend quite a bit of time walking around the lake, and a couple of summers ago I first saw some large fish cruising - doing a circuit of the Mere. It is difficult to say *how* large, but from bow waves etc it gave the impression of being far bigger than anything I was used to. I am used to seeing carp either in the Mere or in any of the ponds around here. and the wake they leave behind them, but this thing was doing rapid circuits of the Mere. So it was obviously quite a large object.

FREEMAN: Did it break the surface at any time? If so can you describe what you saw?

WISNIEWSKI: I was getting the impression of a dark shape below the surface, but while I was watching it didn't actually break the surface of the water. All I had was the impression of a very large fish just below the surface. Usually when fish break the surface - carp or whatever - you get the impression of lots of scales and whatever. However, because on this occasion whatever it was didn't break the surface I didn't get any impression of any sort of scalation.

FREEMAN: Was there any reaction from the birds?

WISNIEWSKI: At the time that I saw the thing it was the summer when we have far fewer birds on site here, and there were therefore far fewer birds on the Mere. What birds were there were probably far more scared of me than of what was below the water.

FREEMAN: Was that the only time that you ever saw it?

WISNIEWSKI: That was the first time that I saw a large fish on the Mere itself. I had repeat experiences in subsequent summers, but this was the best view I ever had of anything so far.

FREEMAN: How long have you been at the Mere now?

WISNIEWSKI: I have worked at Martin Mere for nearly twenty years, but I have only been living on site for about ten years.

FREEMAN: That would suggest that this is something that has come in relatively recently...

WISNIEWSKI: Either something that has come in and grown - because the Mere connects with a river called the Boathouse Sluice, and we pump water from the Boathouse Sluice into the ponds here and quite a lot of small fish we know do come through our pumping system. It's not a screened pumping system which excludes fish. It will allow fish through.

FREEMAN: If I remember rightly there have been records of wels being caught in the area

WISNIEWSKI: There are some records of them being found in fishing ponds in the area. When these ponds were drained for maintenance purposes, I gather small wels were found.

After Richard had interviewed Pat we gave Sarah a briefing on the history of the events at the lake. We told her how we had been trying to find out more about the other wels catfish that had been caught in the area but to no avail whatsoever. The best that we had managed was a telephone conversation a week or so before with a gentleman whose number Tim Matthews had managed to unearth for us. He was a representative of a local angling society and he confirmed that small wels *had* indeed been caught in ponds in the area. However, the only ones of which he was aware were some that had been introduced into gravel pits near Blackpool about twenty years ago, and he had no knowledge of any other wels in west Lancashire.

He did provide us with a momentary twinge of excitement when he told Richard of an incident where a giant pike in the north of England was attacking and killing swans. He promised to hunt out the details for Richard and rang off. Richard was excited by these revelations until he spoke to the man again a few minutes later. It turned out that his source for this report was a novel called *The Pike* by Cliff Temblo.

Richard had actually read this novel which is apparently quite good and tells the story of a pike the size of a Great White Shark on the loose in Lake Windermere. Our informant had, apparently, taken this piece of entertaining hokum at face value and assumed it was true.

Such is life.

Apart from this, however, we had no other reports of wels in the west Lancashire area and had exhausted most of our other avenues of inquiry. All the angling societies mentioned on the Internet as being in that area had invalid telephone numbers and/or email addresses and we had pretty well given up on this stream of our investigations.

Sarah, however, seemed impressed with the sheer scale of the investigations described in this book and asked a lot of intelligent and incisive questions which ended up being printed in what to both Richard and me, was the best piece written about the expedition in any newspaper.

THEY CAME, THEY SAW....

But did they conquer the Mere Monster?

SEARCHING FOR MARTY
Swan eating beast trapped by hunters

SARAH BURDON joins the monster hunters at Martin Mere

The inflatable dinghy lurches through the waters of the Mere, the paddle in Graham's hands disturbing the thick silt on the lake's bottom releasing a rancid odour. My life is also in Graham's hands – this being the potential living quarters of a monster fish up to eight feet long with sharp jaws and a taste for swans.

I sit helplessly at the rear of the dinghy, and as the dark liquid swirls beneath us wonder if we are being watched from below. An eerie calm has settled over the lake which is strangely lacking in signs of birdlife.

Graham suddenly stops paddling and we float freely at the mercy of the water's currents. I watched countless bubbles explode on the surface, my mind playing tricks on me as I envisage a circular motion of bubbles as a giant predator sizes up his prey.

Sweat trickles down my face from the relentless glare of the sun and my life jacket feels claustrophobic as we watch…… and wait. But there is nothing, only the sound of the breeze through the flora surrounding the lake. Still I breathe a sigh of relief when we eventually head back to shore looking forward to having my feet once again planted firmly on solid ground. A hub of activity greets us back on terra firma as a team from Sky News conducts a live broadcast on the Martin Mere expedition.

The rest of the four man team are inside the New Raines Observatory where they have spent the last three nights eating, sleeping and breathing the mysterious underwater activity at the Mere.

Together they look a motley crew – cryptozoologist Richard Freeman, Graham Inglis and John Fuller dressed in combats while director Jon Downes, a genial hulk of a figure, presides over the proceedings from his laptop on which he is writing a journal of the expedition's progress.

232

The musings – already 70,000 words long – will form a book to be produced around Christmas time. This is the final day of the investigations and tomorrow the team from the Centre for Fortean Zoology will be heading back to Exeter.

SIGHTINGS

Following three visual sightings and four positive readings on the fish-finder equipment, Jon is happy with the expedition's outcome. "The trip has been successful – we have had three visual and four sonar sightings of what appears to be a wels catfish – the largest freshwater fish in England. The biggest fish around here are carp and they are a green, brown and golden colour. The fish we have spotted was an oily black – a minimum of four feet long and a maximum of eight feet – but the wrong shape for an eel. This fish, as wide as a barrel, can slither across muddy ground and grows very slowly – this one could be a hundred years of age. It would then be the oldest fish in England as well as the biggest."

Richard, who has spotted the fish three times, said:

"It was 9.30 pm on Thursday, the day we arrived. I was walking along here when I saw the back of a large fish with black and oily skin. There were no scales, but it had a tiny crest down the length of its back. It made a big disturbance in the water. On Friday morning I saw it again just below the surface – it would take a man to make the kind of commotion going on down there"

The event has attracted a lot of attention both from the media and from people visiting Martin Mere. Jon said:

"You are my 38th interviewer in the last two days! But I've enjoyed it – I'm a complete media slut. It's been brilliant. We have received such a positive response from everyone. I have had children running up to me and saying 'Excuse me mister, how can I become a monster hunter?' That's great – it beats an obsession with PlayStation or television".

The team survives exclusively on fund-raising, so is on the look out for sponsorship or donations. Anyone willing to help can contact Jon on 01392 424811.

The team also needs donations of equipment to help on their expeditions.

There is a very real need for a tea urn at present.

Just then a grey haired couple arrived on the scene. They were David and Estelle Walsh, the people who had first reported the sighting of the swan being attacked back in February. It was essential that we talk to them, but just as we were setting up for an interview a truck arrived containing a crew from *Sky News*, who then proceeded to take over the whole day and completely disrupt most of the activities that we had planned for the rest of the day.

As I introduced Pat and myself to the news reporter, whom it has to be said, was perfectly affable and charming in a slightly detached sort of way, Richard and Graham managed to complete an interview with the Walsh's. Estelle did the talking whereas her husband, backed up everything she said with gestures and supportive grunts of agreement.

FREEMAN: Basically could you tell us what you saw, what happened and what time of year?

WALSH: It was in early February. We had been coming for several nights for several weeks and the Mere was covered in swans, geese and other waterfowl. They suddenly lifted from the surface of the Mere in front of Swan Link until the Mere was empty. This had happened on several occasions.

This time the Mere cleared but it left one swan in distress somewhere near the centre of the Mere. The swan was screaming and fighting. It was kicking screaming and fighting against something that had been holding it underwater and it took about half an hour for it to reach the bank of the lake over by Raines Hide here.

It managed after a while to reach dry land and it just collapsed, exhausted. As it relaxed it was then pulled back into the water, and it started to struggle and kick and scream for another half an hour until it managed to get back onto dry land and pull itself far enough out of the water to be safe. It was exhausted. At this point Pat Wisniewski came in.
FREEMAN: How badly damaged was the swan?

WALSH: At that point it just looked tired, exhausted and frightened and Pat decided that he was just going to leave it there to recover there. But the following morning we were back at the Mere and there was a swan reluctant to go into the water at the edge of the Mere, and it had a piece - like a semi circle - missing out of its wing. This wasn't what we were used to seeing in crash injuries and other injuries with swans and so it led us to make the assumption that it was the same bird.

FREEMAN: Is the injured bird still here now?

WALSH: It's not on the Mere but it's still on the reserve. We saw it on Wednesday. The two birds that are on the Mere now are ringed birds and this one wasn't ringed.

FREEMAN: So it was so badly damaged it didn't migrate?

WALSH: No

FREEMAN: And you didn't see whatever it was beneath the water?

WALSH: No

There is something about the mass media which seduces you. I have been working with them for long enough that one would have thought that I would be immune from their blandishments by now. However, I am as susceptible to their techniques as anyone else and the events of Sunday bore this out horribly.

Now I don't want to be one of these people who bore on at length about money matters. The CFZ is desperately underfunded, but then again considering the nature of what we do this is hardly surprising. I am determined to establish the Centre for Fortean Zoology in the public mind as a bona fide research and animal welfare organisation. In order to establish our credentials I have had to turn my back on some of the tackier and more crass avenues open to us in order to raise funds. I never became a cryptozoologist in order to make money, and I am quite happy with the amount of money that I make for myself. My personal income is sufficient for my own needs, and I am happy to continue financing the CFZ for as long as is necessary.

However, this investigation has cost me personally something over two and a half grand, and I am unlikely to recoup any of it, in the short term at least. I was hoping to use the medium of *Sky News* to appeal for potential sponsors. If the CFZ is going to expanding the areas that we wish to over the next few years, alternative forms of funding are essential and this seemed to be a perfect opportunity to make a few quid for the cause.

I told the *Sky News* reporter what I wanted to do and he appeared to agree with my request. Now, I want to stress that he NEVER agreed to let me make an appeal on live television, but I was led to believe all through that long afternoon that I would eventually be allowed to make an appeal on television. The first broadcast went out at lunchtime and featured David and Estelle Walsh and Pat Wisniewski.

In the background were Graham and Sarah Burdon paddling around in the boat in the middle distance. Almost without us realising it Sarah had become an integral part of the CFZ investigations team, for a few hours at least. The rest of us sat on the grass outside the hide, watching the TV crew at work. The presenter marched up and down the lakeside talking to the camera, and as the two in the boat paddled up and down behind him, a large solitary pink footed gander marched up and down the lakeside at the presenter's heels.

There was no doubt whatsoever, as to whom, in the gander's mind at least, was the boss of the place. We saw him later in the evening patrolling around the outside perimeter fence of the wildfowl nursery area, making sure the mergansers and demoiselle cranes were tucked up for the night before he retired to whatever bolthole the regal fowl had carved out for himself.

When the broadcast was over, the TV crew (and the gander) disappeared, and as I still had vain thoughts that I was going to be able to appeal to the Great British Public through the medium of Murdoch's media, I spent much of the afternoon hanging around the base camp at the New Raines Observatory. The TV team turned up after a while and without asking co-opted half my team into their activities almost without me being aware of the fact.

Graham was sitting on the edge of the water like a garden gnome. He was brandishing a long piece of bamboo onto the end of which he had tied the sonar fishfinder transducer lead which was dangling into the water. He managed to fake up some fish sightings on the sonar receiver and the TV company filmed the "evidence" They also filmed Richard and John bustling around and me sat at the laptop in the New Raines Observatory typing in the preliminary results of the investigation.

The afternoon wasn't entirely wasted. Pat and I had several long talks and began to hatch the basics of a plan for a return trip to the lake during the summer of 2003. We even began to discuss a strategy for catching the creature. The main problem with catching the 'Monster of the Mere' had always been, what we were to do with the fish once we had managed to catch it. Now, pat came up with the germ of an idea. The martin Mere

236

site is divided into two parts; one a zoo with a collection of exotic water-fowl and the other the natural wilderness of the Mere. Under the terms of the 1980 Wildlife and Countryside Act it would be illegal to re-release any non native creature into the wild. We are hypothetically facing the same problem with regards to our plans to carry out an in depth scientific study of the big cats which live in various parts of the United Kingdom. Our consultant ecologist Alayne Matheison, who carried out valuable work with lion and leopard populations in Zimbabwe, is not able to set up box snares on Bodmin moor as she did in southern Africa. Because if we capture a big cat, we can't let it go again, and what the heck are we *actually* going to do with it?

Until now we had faced the same problem with the 'Monster of the Mere'. Killing the creature was completely out of the question, and none of us felt particularly good about the idea of rehousing it in an indoor tank at an aquarium, zoo or Sealife centre, even supposing that we could find one willing and able to take on such an enormous new exhibit.

Pat came up with a possible solution. There was a large pool in the zoo side of the reserve, and there was very little in it. It had a UV filter and various other mod cons from a fishy point of view. If we were able to catch the fish on our return to the lake a year hence, and that is still an enormous IF, there is no doubt that Marty would make an extremely impressive tourist attraction and fund-raiser for the WWT reserve.

We spent much of the afternoon chatting happily about the methodology that would be required in trying to catch it. I also had an extremely important PR task in hand. The enormous number of radio and newspaper interviews that I had carried out over the previous few days had paid dividends, and there were far more visitors to the reserve than usual. What I was not prepared for, however, were the sheer numbers of people who just wanted to come back to the New Raines Observatory and talk to the team, particularly me and Richard, about our hunt.

What was particularly encouraging is that so many were children and that they showed such an interest in what we were doing. I spent quite a large portion of the afternoon talking to kids about monsters, and was rewarded by several little boys (and one little girl) telling me that they had now decided to be cryptozoologists when they grow up. I hope that I may have planted a few seeds in a few fertile minds that afternoon, and sincerely pray that the next generation of cryptoinvestigative people will not have the problems with funding and public acceptance that have so beset the CFZ during our first decade of existence.

The blokes from *Sky News* would turn up at intervals. It seemed that whenever we asked, the time of the second live news broadcast would have been put back and before we knew it most of the afternoon was wasted. This was particularly annoying from my point of view because I was hoping that we would be able to complete the mapping operation on the Sunday afternoon, as well as take some depth soundings on parts of the lake that we hadn't yet explored. However, we were stuck at base most of the afternoon because every time we would move to do something constructive for a change, the *Sky News* people would arrive as if possessed by some arcane sixth sense and would ask us to stay "just for another quarter of an hour" so as to be on hand for an imminent broadcast, which always then never materialised.

By four thirty I had really had enough. So had Graham, and he disappeared off to the far shore of the lake to continue his mapping activities while I stayed at base camp doing my best to answer a battery of increasingly repetitive questions from newspaper people. If I sound churlish then I really don't mean to. I have no problems at all with meeting with and talking to the public. That is a major part of my job. The CFZ only exists because of public supports, so in many ways, especially when we are in the field, we are public property. If members of the public want to ask me about the Loch Ness Monster all afternoon, it is their right and it is my duty and indeed my pleasure to answer them.

The media are a different thing though. Some newspapers who had already covered the story in some length wanted more, and the editors had obviously sent their reporters in search of another follow up story about the expedition's progress. Well the sad truth is that nothing of note had happened for twenty four hours or more by this point, and although we were still achieving a string of mission objectives they weren't the sort that filled newspaper editors with any real delight.

The following uncredited news story turned up in my e-mail inbox that afternoon:

Sunday 28 July 2002

Monster hunters spot giant fish.
Wels catfish have been known to reach 16ft in length. A
team of paranormal investigators from Exeter believe they
may have identified a giant creature being held responsible
for attacks on swans in Lancashire. Sightings of the crea-
ture, dubbed the Monster of Martin Mere, were reported ear-
lier this year by visitors to the bird reserve at Bur-
scough, in West Lancashire

Now a four-man team from the Exeter-based Centre for
Fortean Zoology say they have seen a "very big fish" during
a 24-hour watch at the Wildfowl and Wetlands Trust beauty
spot near Ormskirk. Organisation members, who claim to have
pursued vampires in Mexico, dragons in Thailand and skunk
apes in Florida, have launched a four-day operation in an
attempt to crack the mystery of the fish.

They are spending four days at Martin Mere using infrared
cameras, military-style night lights and "fish finder" so-
nar equipment in a bid to find out more about the mystery
beast. Director Jonathan Downes, who said he was somewhat
sceptical when he first heard reports of the fish, believes
the monster may be a Wels catfish. Native to mainland
Europe and introduced into parts of the UK in the late 19th
Century, Mr Downes said the Wels catfish is the largest
freshwater fish in the world and can reach a length of
16ft.

More modestly-sized specimens have been found in the area
near Martin Mere and with a lifespan of up to 100 years,
the monster could date back to Queen Victoria's reign. "If
it is a Wels, it is almost certainly a British record,"
added Mr Downes.

The "eight-foot fish" was first spotted yesterday by team
member and qualified cryptozoologist Richard Freeman who is
hoping to capture the creature on film. He said: "I have
seen something black and shiny snaking around in the water
in almost the same place as the original sighting several
months ago. It certainly looked like a Wels catfish.

"However we will be carrying out further investigations
over the weekend in hope of obtaining photographic proof".
Mr Freeman said the fish had no scales, had a "rubbery" ap-
pearance, was oily-black in colour and moved quickly
through the water. "I can't say for sure that it was a Wels
catfish. But if a pike had attacked the swans there would
have been wounds. This thing seems to come up underneath
and drag its prey down under the water."

Reports of a larger-than-life creature living in the 17-acre lake were first voiced four years ago and the Martin Mere monster has since become a talking point among people living near the 380-acre reserve which regularly attracts Hooper [sic] and Bewick swans. artbell.com Abby Normal news....

Just as I was getting terminally bored with answering the same questions over and over again the sound of the steady rolling tread of *Das Geschlecht der Matthews* grew louder as the entire clan burst into the tiny hide. Tim led the way beaming away in his most subversive manner, closely followed by Alexandra who makes more noise than an entire division of panzers, which is surprising for someone as angelic as she is who ain't even three yet! *"Uncle Richard!!!"* she screamed, ignoring me entirely and literally jumping into Richard's arms in search of cuddles and chocolate.

In five minutes that Matthews brigade had annexed the hide as surely as if it had historically been part of their little house in suburban Southport. *"Guten tag mein herr"* I said to Matthews, trying to discern whether the rumours were true and that he was trying to grow a little toothbrush moustache. *"Enough of the German stuff already"* he said with a grin *"tell me about this bloody catfish"*.

Over the next ten minutes we quaffed tea, sat on the grass under the beating July sun, and I filled Tim and Lynda in on the situation so far. As we talked and planned our next move, and continued to wait for the Sky News people to actually make up their mind about what they were going to do I mused on the sheer surreality of the scene. If only O'Hara could see us now, I chuckled to myself. A fat bloke with the sort of haircut which if you had read *At war with the universe* would have made one assume that Tim would have immediately had me arrested as a subversive, sitting with an alleged neo-fascist government agent (who was of course nothing of the sort), while a plumpish, balding Goth in military fatigues played with the blonde, blue eyed daughter of the alleged neo-fascist and a black guy in uniform made us all tea.

It was a peculiarly domestic scene, and one which would not have been at all out of place in one of the chronicles of middle class English childhood so beloved of John and me.

The *Sky News* people returned and began to issue their commands for the live broadcast. As readers of this chronicle so far will be unsurprised to find out they decided in their infinite wisdom not to interview me, but instead they wanted to talk to Richard about his sighting while Tim and Graham paddled around in the middle distance doing absolutely nothing of concrete importance in the boat. By this time I was not only rather cross, but getting horribly tired as well and decided to give the whole affair up as a bad job.

FROM GRAHAM'S LOG:

Hopefully the recorded *Sky News* item is going out, as I type. Earlier, the *Sky News* reporter, Trevor someone, interviewed sightings witnesses while I and Sarah Burdon (Champion Newspapers) paddled the dingy in the background - in "ever-decreasing circles", as Teen, who saw the live report earlier, put it.

Then they recorded bits of me using the fish finder (I faked up a fish contact). And Trevor's just told me they're doing a live piece around 5.30... so I'll get out on the water again!!

Anyway, most of the last day (Sunday) was spent entertaining the *Sky News* mob, which annoyed Jon because he'd been under the impression that he'd be allowed to do a piece including an appeal for donations... whereas my impression from what Teen (Tina Askew - a friend of mine and Graham's) told me was that what they really wanted was lightweight Sunday News type stuff. And interestingly, Lisa (my daughter) was surprised to see my video of the lake: she hadn't realised it was so nice-looking. From the Sky News coverage, she'd got the impression it was just a normal sort of pond. So apparently they didn't even bother to get a decent shot of the scenery.

Teen's gonna send me the tape she did of their coverage, and I'll copy it to Sarah (the lass who accompanied me in the dingy).

By this time, Sarah had gone home, and John and I were left sitting on the edge of the shore fulminating about the perils of getting involved with the mass media. Never again, I thought, but like a not very devout Catholic in confessional or an avowed alcoholic in an AA meeting that he didn't want to go, to I knew damn well that next time the mass media came calling, I would be sniffing around them like a stray dog round a bitch in heat.

As I had told Sarah earlier, I have become a media tart. I do every possible TV show offered me, not because of any desire for self aggrandisement or even because I enjoy it, but because I am realistic to realise that the media are a very useful weapon in the armoury of the CFZ. In fact they are our most useful weapon, because they are our only conduit to the general public, and without the support of the general public the CFZ will never be able to survive.

The broadcast seemed to take an inordinate length of time to complete, but finally it was over and we could get on with what remained of the rest of the day. Because of the sheer amount of time that we had wasted faffing about with the team from Sky News (who were, by the way, extremely gracious as they upped and went), we had lost the chance to do some of the sonar depth mapping that we had planned for the Sunday afternoon. so whilst the rest of us began to pack up the gear because we had a long drive ahead of us on the following day and we wanted to do as much as we could the night before, and said goodbye to the Matthews clan (because it was time for baby Freya's tea) Graham disappeared to the other side of the lake to finish his land-based mapping activities. We were now not going to be able to finish the underwater mapping of the lake that we had planned, purely because of the activities of the media folk, but we were as sure as hell going to be able to finish the mapping of the shoreline and islands.

Richard, John and me decided to have one last bash at trying to bait out the creature from its lair deep in Rebecca's Channel, but as we understood the beast to be primarily a crepuscular feeder, we had a couple of hours to go before it was appropriate to commence baiting activities, and so after packing up as much of the equipment as we could we took a well earned rest..

The relationship between me and Graham had deteriorated badly during the Martin Mere expedition. We have been friends for nearly fifteen years, had a casual work relationship for a decade, and been close colleagues for the last six. However during these few days at Martin Mere the atmosphere between us was horribly redolent of the last few weeks before my wife and I separated.

With horrible irony today was the sixth anniversary - to the day - of the worst day of my life. On July 28th 1996, after weeks of bickering my wife and I had an argument over something relatively trivial. I fell asleep, and when I woke up she and my daughter were gone. There was a vitriolic note pinned to the front door, and the only other communications I have had since have been through our solicitors. Lisa made contact a year or two later, and eventually moved back in with me, but I never spoke to my wife again.

Despite my deteriorating relationship with Graham, I was determined never to go through that sort of scenario again. It was clear that our relationship would have to change, and it was also clear that we may not be able to work together in the same way as we had done, but I was determined that whatever happened to our relationship it was going to happen in a civilised way.

The main bone of contention between us is was Graham refers to as "head space". I am a reasonably gregarious person whereas Graham is not. I also work pretty well within a team situation where Graham is most definitely a lone wolf. Over the years, as the CFZ has become more professional and successful this has become more of an issue. When Graham first became the Deputy Director of the CFZ in the immediate aftermath of my wife's and my separation, I was wallowing in a mire of alcohol and lachrymose self pity. I would not have got through this period if it hadn't been for Graham, but the protracted sessions of alcohol abuse that followed were too much for me and I eventually cut down severely on my drinking. Graham, however, enjoyed the anarchic and booze fuelled madness of the mid period CFZ. These are the days portrayed in my semi-novel *The Blackdown Mystery* and in a highly scurrilous book currently being written by my friend and colleague Nick Redfern.

Together we carved a boozy swathe across England, and even across Central America. Our adventures were many and our exploits memorable. (Or at least they would have been memorable if we hadn't been too drunk to remember some of them). On a cloud of loud rock music and cheap whiskey we investigated everything from animal mutilations to UFOs, and from big cats to vampires and in doing so carved out a significant reputation for ourselves.

However, partly for health reasons, partly for money reasons and partly because I had really had enough of the repeated embarrassment of waking up in the morning not knowing what the hell I had done the night, before I seriously cut down on my drinking. My life had become similar to that outlined by Loudon Wainwright III in *Wine with Dinner:*

"Don't know who I insulted/Maybe it was you,
The drinks I had resulted/ In a tirade or two,
went down town to get in trouble/I accomplished that..."

When I sobered up I realised what had happened. During our protracted lost weekend (which had lasted approximately two and a half years), although I had made myself a reputation - and astonishingly not one of a drunken buffoon - I had made and spent an awful lot of money and the business affairs of the CFZ were in a parlous state. We had lost a lot of subscribers, purely because our administrative procedures had been so shoddy that book orders and resubscriptions had remained unfilled and some people were just not getting what they paid for whereas others were getting their orders filled twice or even three times over.

Something had to change, and after a serious bout of mental and emotional ill health at the end of 2000 and the beginning of 2001 I began to try and claw my way back towards some semblance of normality. Over the last two years I have gone from being a shambling drunk to being the head of an organisation whose presentation and business methods are now as good as its research and publications. This has eventually led to us having Colonel John Blashford-Snell OBE as our Hon. Life President, and the imminent likelihood of proper charitable status.

Putting our/my house in order has been a difficult and painful task and is by no means finished yet. It has involved not only stopping drinking but pruning the gaggle of freeloaders who for years hung around the periphery of the CFZ making it party central rather than the headquarters of the world's premier cryptozoological organisation. For the last few months of 2000 my house/the CFZ (for they are one and the same), had seemed to have become a drop in centre for every drug addict, transsexual, pervert, Satanist and general loony in Exeter. Somehow these weirdoes had wormed their way into my household where they stayed drinking my wine, eating my food and bleeding me dry. At the time Graham was so disgusted with these people that he just kept his distance not realising that the situation was totally out of my control and that I hated these wasters being there as much as he did but that I just didn't have the emotional tools to get rid of them.

By the spring of 2001 they had all gone, and eventually *some* people who had been quite close friends of mine had gone too. I became aware, sadly and with great embarrassment, quite how many of my friendships had been based primarily on substance abuse and a dysfunctional life-style, and as I stopped my self destructive behaviours it was sad to see how much of my social life disappeared along with it.

I filled the void with work, and by the summer of 2002 my life is happier and more successful than it has been in years. However, not all the changes suited Graham. Like me he was only too pleased to see the back of the freeloaders, but although he saw the need for greater efficiency he disliked the more structured regime at the CFZ. Graham is one of nature's anarchists (in the true sense of the world), and - despite what I may have said over the years in song lyrics, magazine articles and even books - I'm not. Because of my mental health problems I need a far greater degree of structure in my life than many people, and sadly as Captain of the good ship CFZ, if I am to fulfil my duties in an acceptable manner the organisation needs a far greater degree of structure than many people involved (especially Graham) would have liked.

He *hated* the uniform idea for example, and we fought tooth and nail over the concept until we grudgingly agreed on a compromise which sat-isfied neither of us. He saw it as pointless regimentation and even as self-aggrandisement on my behalf, where as I saw it as a sensible way of pre-senting a homogenous front to a world who really didn't understand what we were trying to do. He hated having to wear life jackets, seeing them as a manifestation of what he called "The Nanny State" - a dull and grey Britain where children are no longer allowed to make daisy chains (unhygienic), or play conkers (dangerous). As I had been "messing about in boats" on and off since I was six and always wore life jackets when doing so, I had no such problems, but we had some bitter arguments over the issue. However the thing which upset Graham most was the sheer lack of personal "head space" that he could have during the expedition and the time immediately leading up to it.

```
FROM GRAHAM'S LOG:

Now it's 1300 on Fri (Day 2) and I've had no chill time
worth spit since Monday, I reckon. Let's backtrack and work
it out...

Thurs started 5am and was the driving day, settling in,
etc, and scrambling the dingy when RF had his initial
sighting. Also, after all our talk about getting a rope
```

```
across an expanse of water, we finally did it. However, the
dingy wasn't required for that. The rope runs from the
northern tip of Turner Island, over to the Swan Box, a dis-
tance of around 300 ft.

Continuing the backtrack:

Wed was the final day of mission preps, when I didn't get
home til about 10pm and had to get up again at 5am.

Tue was the meltdowns day and my paltry amount of time on
my own afterwards was stress recovery, not chill-time or
headspace.

Yeah, I can't remember what I did Mon but hopefully I en-
joyed some of it. Anyway, that was a long time ago now.
```

Graham's levels of resentment were rising rapidly and there was nothing that I could do about it. I had to keep up the level of work necessary to achieve our targets or else all the time and expense that had been lavished on the project would have been wasted. The simple fact that we had been messed around for a whole afternoon by the team from *Sky News* meant that we had not been able to fulfil some of our aims. If the team hadn't pulled together throughout the weekend, and for the week or so previous to the expedition (and indeed the few days immediately after it as well), then absolutely nothing would have been achieved and it would all have been an ignominious waste of time and money. Richard and I are driven to do what we do and therefore accept these privations as necessary obstacles upon our route towards cryptozoological nirvana. It seems (although he has not been working with us for very long) that John Fuller does too. After years not knowing what to do with his life he has embraced Cryptozoology with a vengeance at the age of forty and is now as enthusiastic as Richard or me. However Graham is not driven in anything like the same way as we are. It has not been his lifelong ambition to hunt monsters and discover new species, and he finds the extra pressures surrounding an expedition and the preparations for it to be irksome almost to the point of being unbearable.

Another bugbear was the issue of the "curfew". Graham likes "doing his own thing" and being answerable to no-one, whereas I have the belief that in the field I am responsible for the welfare of my team and also for the welfare of each person within it. Indeed if one of my team members were to be injured, or killed in the field and a court of enquiry was to find that I was responsible, I could find myself facing charges up to and

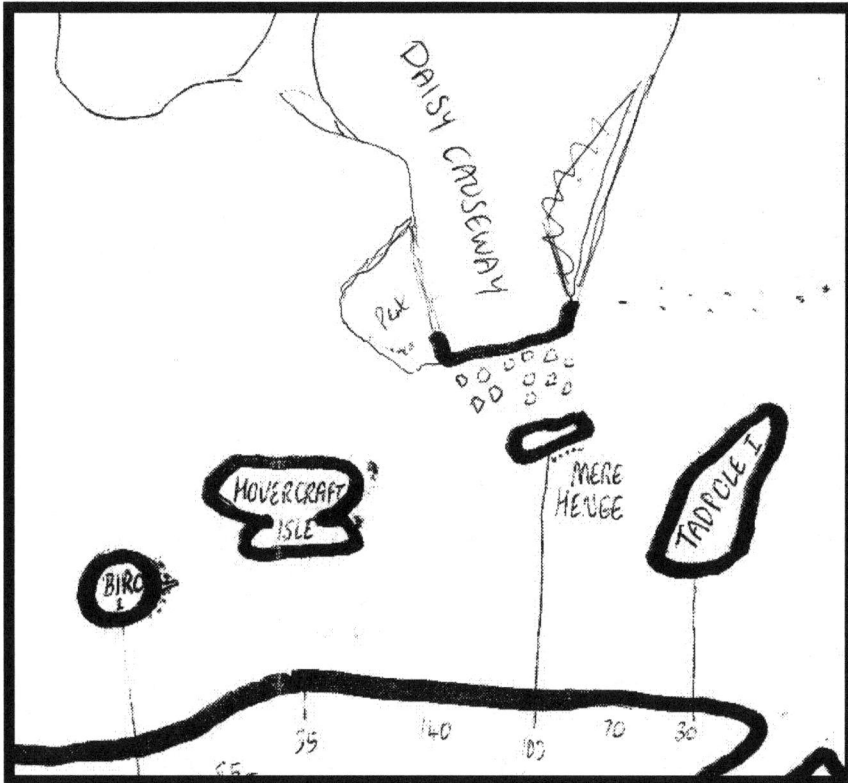

At one point, I had the idea of photographing each isle from the adjoining isle, using the digital stills camera held as high above my head as possible, and later coalescing that information into the overall cartographical stew. Leaving aside the administrative problems of logging which way the camera was pointed for each shot, I soon realised that a photo taken from 8½ ft above the ground was of little more use than my own eye-level observations 6 ft above the ground. Some crop circle photographers elevate their camera on a pole, 30 or more feet in the air, and take pics, and I could have done with a similar facility. Now, there's an idea for the future. In its absence, however, I just kept on wading and plodding around, and continuing with my jigsaw puzzle.

I gradually fleshed out the detail of the shape of each snadbank or islet, and re-estimated the distances between each one, and stuck each element onto the master map. Eventually the above-water mapping was complete, and we could turn to assessing the depths of the various channels, which could easily be done by sonar from the dingy.

including culpable manslaughter.

FROM GRAHAM'S LOG:

Chill time last night was severely contaminated when I announced I was off for a wander (instead of just disappearing) and Jon suggested search parties if I hadn't returned
in an hour. I said, "I'm not gonna submit to a curfew!" and
Jon denied it was a curfew but just said they'd be "getting
worried" if I hadn't returned by then. Out by the southern
shore of the lake, the vibes were very different to the
previous night. I had this curfew cut-off point hanging
over me, and sullenly returned after a while.

I'd joined the group and was contemplating being social,
but then he asked me, "What did you do with the beer cans?"
I just snapped, "I threw 'em in the sodding Mere, didn't
I?!", which amused RF. It's a severe mistake trying to mix
work and leisure. A shame, when visiting such a nice area,
but when I'm sitting out at the far end of the lake, and
can hear Jon's cackling from over 150 m away, and am mull-
ing over the Curfew Issue, then I'm conscious that really
I'm still in a bloody work environment...

The night following the Curfew Issue, ie **Sunday**, was our
final night there and when I announced I was going off for
a wander, Jon showed little interest, which was rather a
relief. This enabled decent chill time comparable to that
of Friday night (the "search party" night).

I dropped in to see the others back in the Raines hide from
time to time and once to stash the car keys (so I wouldn't
lose them) and was somewhat embarrassed to overhear Jon
saying, "Perhaps Graham's right, and I don't understand his
headspace needs." So I discreetly backed off and went back
out to the lake, thinking that if Jon was bothering to make
an attempt to get his head around the matter, I didn't want
to interrupt, or eavesdrop. However, when I commented on
this to John, the next day, he said Jon's theme had been
more on how he, Jon, finds it difficult to relax. Which
he's conceded often enough to me anyway, so no new under-
standings there, after all.

By Sunday night the relationship between us was so bad that we were
hardly speaking to each other. Clearly we would both have to do a lot of
work if we were ever to venture out into the field together again. At
about six o'clock Graham and John drove into Burscough for supplies
and returned with some of the most revolting fast food I have ever eaten.
I couldn't eat most of it and so I decided that as I was supposed to be on

a diet anyway I would do without and wandered off to look at some chickens.

For some reason I have always had a mild obsession with poultry. I like the noises that they make and the earnest way that they scratch around in the dust looking for worms, other small invertebrates or scraps of organic detritus. Sad, because I could see myself imminently having to make a judgement call between my best friend and the future of the Centre for Fortean Zoology I wandered over to a far flung part of the nature reserve where I sat on a log, smoked a cigarette (blithely oblivious of the fact that I had given up at Christmas), and watched a small flock of chickens bustling around living a poultry orientated lifestyle.

My musings were interrupted by Richard and John who were perfectly aware of the burgeoning problems between Graham and me and were powerless to do anything about it. Richard decided (in his infinite wisdom) that the only way to cheer me up was to take me to see some hens. He was perfectly right and by the time we had spent a pleasant half hour wandering past the wildfowl nurseries where there were not only various types of wonderful poultry but a small group of young mergansers, looking almost like prehistoric birds with their cruel reptilian beaks and their wild staring eyes, and even a pair of Asian demoiselle cranes - a species I remember from my youth in the far east, I had cheered up and was ready to resume investigations on the last night of our expedition.

We left Graham to his own devices, and made our way to FishBase1 where we sat on the cool grass and scanned the water of Rebecca's Channel and the far side of Biro Island. We ground baited intensely for about an hour, and at about 8.00 pm Richard carried out our last revolting experiment. He fetched the dead lapwing which he had found the previous evening and stuffed its body cavity with a revolting mixture of sardines and mashed up corned beef.. He tied it to a long length of yarn, and using the skills which apparently had made him quite a notable bowler in his school cricket matched lobbed the unfortunate (and thankfully very dead) bird into the deep water and pulled it backwards and forwards across the water in an attempt to tempt "Marty" to the surface.

Unfortunately absolutely damn all happened. In my head I could hear the twin voices of my Army Chaplain brother intoning *"cans't thou draw out leviathan with a hook?"* and my old mentor, the notorious Irish wizard Tony "Doc" Shiels ranting through his bristling grey beard: *"If you*

249

want to see a lake monster you have to invoke the focker. Ther's no point in just throwing a shagging lapwing at the feckin' thing". I had a sneaking suspicion that he was probably right (even if he was speaking to me in a voice that only I could hear.

Just then there were two enormous splashes at the other side of the lake, and several hundred yards away, through the gathering half light we could see enormous ripples where something extremely large had broken the surface of the water. There were no waterfowl whatsoever in that portion of the lake. Had they been frightened off by the commotion? Even more exciting a possibility: had they vamoosed after one of their number had been dragged to a watery doom by the Monster of the Mere? It was certainly a possibility and one definitely worth investigating.

Making sure that their walkie talkies were switched on and that they were both wearing life jackets because it was getting dark and I did not want any member of my team (who would actually listen to my instructions on the matter) wandering around near water at this time of night without one, I asked John and Richard to investigate the splashes and any other disturbances on that side of the lake and I stayed at what had once been FishBase1 to continue dragging the lapwing through the water and hope that I would eventually get a sighting of the monstrous fish.

After a while I sat back down on the grassy bank and let the lapwing on the string float out onto the dark waters of the lake. Maybe the fish would be more attracted by a stationary target than a moving one. I broke all my own rules about drinking alcohol in the field, reached in my pocket and got out a half bottle of scotch. Taking a healthy swig, and lighting another cigarette (with a silent vow that I would once again have to give up when I got home), I surveyed the silent waters of Rebecca's Channel and waited for something to happen. On the far side of the lake I could see John and Richard peering into the water. Intermittently my reverie would be interrupted by them radioing in to tell me their progress. As they had a big carrier bag of ground bait with them (and also because I fancied a little bit of headspace of my own) I suggested that they stay there and ground bait for thirty minutes or so. Looking back towards the camp I could see Graham, video camera in hand, apparently trying to film a traffic bollard which for some reason had been left by the lakeside. Shrugging my shoulders I leaned back on the soft turf, took a drag on my cigarette and began to daydream about the lady after whom Rebecca's Channel had been named.

Would our relationship survive the test of time and the separation of an entire ocean? Was I just deluding myself into having feelings for her? Could someone possibly build up a viable relationship with someone over the Internet? I would always have said no, but here I was, emotionally involved with a girl I had never met and only spoken to twice on the telephone. Surely in the absurd history of events which has been my life this was the most absurd thing that I had ever done? All I knew was that at the moment at least she was making me happy, and I seemed to be making her happy which is far more than could be said about most of the relationships in which I had been over the years. Only time would tell what was going to become of the affair (in my minds eye we had met, fallen in love IRL and eloped to live together in a castle in the air), but I knew that I hadn't spoken or had electronic contact with her for four days now and that I was getting withdrawal symptoms.

As the light finally faded a radio message came in from Richard and John. It was too dark to see anything they said. They had managed a few short sightings of the dorsal fin of what seemed to be a huge carp, standing proudly brown and gold above the water surface, but nothing else. It was time, they thought, to return to base, and they were coming back to FishBase1 to help move the rest of our equipment back to the New Raines Observatory.

I agreed, and in the last few minutes before they returned, a fat, bearded man with the cares of the world upon his shoulders daydreamed about his cyber lover and watched in awe as an enormous wake slowly cut across the surface of the lake about thirty yards in front of me. This was an intensely private moment. I could see that something huge was cutting an incisive swathe across the water's surface. At last (if it were not wishful thinking and I didn't think that it was), s piece of firm evidence for Marty's existence with my own eyes. I may not have seen the giant fish itself, but in the few minutes before night finally fell I had finally discovered for myself that something extremely large, and extremely alien, *was* indeed lurking beneath the waters of Martin Mere.

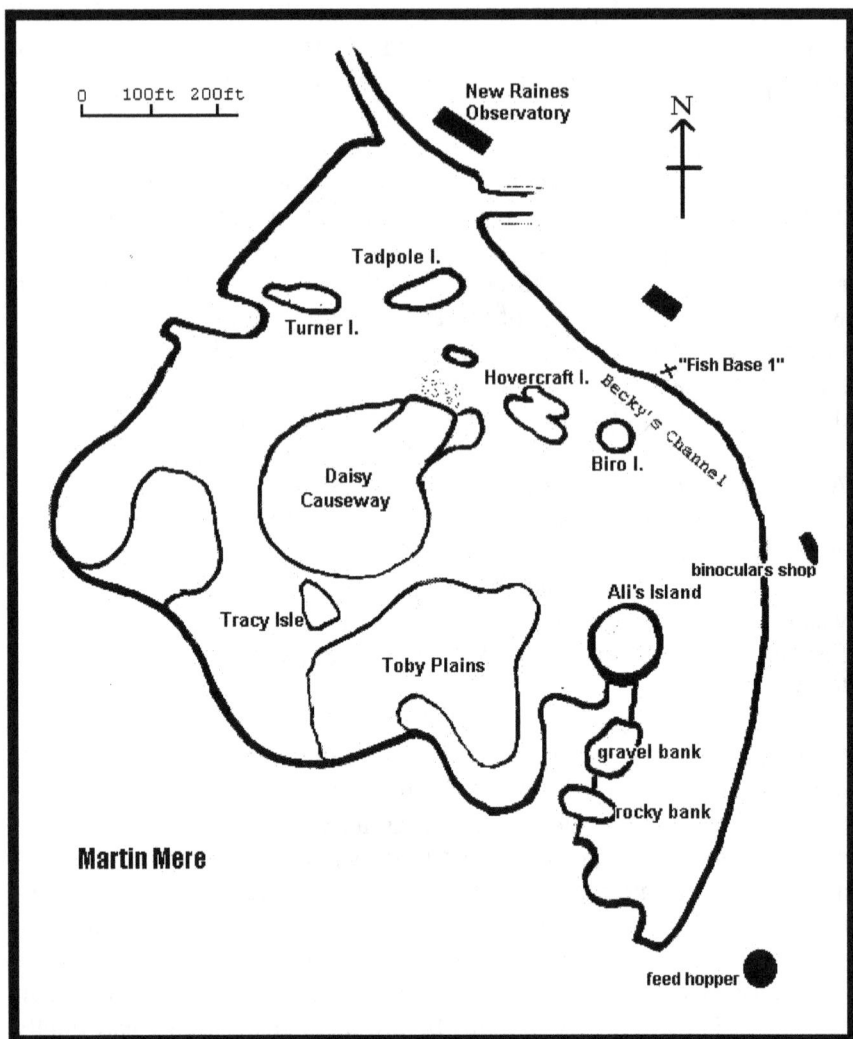

Graham's completed map

Monday 29th July

There is not much else to say. We awoke reasonably early and had the entire camp dis-assembled and packed into the two cars by 8.30. We were supposed to be seeing someone from the *Liverpool Echo* at 11.00 but as he hadn't arrived we decided to go. There was one final visit from a journalist - a young lady called Katy from the local paper who wrote a eulogistic piece about us which portrayed us in a far more heroic light than I think we deserved.

FROM GRAHAM'S LOG:

The drive back was a lot more comfortable than the journey up. Okay, I had a slight hangover, but that's nothing unusual. I'd got a decent night's sleep in the Volvo, parked out in the public area near the second picnic table, and left the windows and tailgate open to ensure plenty of refreshing night air, so I woke up feeling quite cheerful.

We set off at midday and this time, I split the journey into chunks of around 50 miles. I think Jon went for 80 mile chunks, as on the way up, altho' I haven't looked at that particularly closely. The Volvo cruises happily at 70 and I stayed with that most of the time.

Just before we left, Pat came to say goodbye. By this time we had agreed pretty well for definite that we would be returning during the summer of 2003 to try and net the beast which we now all knew was there. So it was not goodbye, rather *au revoir*. However all partings are tinged with sorrow and after such a highly successful expedition, which despite personal and logistical problems had been idyllic for the most part, it was particularly sad to leave. We shook hands firmly, and as he stood by the gate of the perimeter fence waving farewell to us, the tiny convoy - the jag in the lead and the battered old CFZ Volvo bringing up the rear - drove off towards the motorway and eventually to Exeter.

Our adventure was over.

HAIL AND FAREWELL

But of course it wasn't over at all.

As we drove out of the gate, our pleasure at the successful completion of our mission tinged with the sadness of leaving such a fantastic place, and the people who lived there, Richard and I had no idea that scarcely 48 hours later we would be back.

We had an uneventful drive home and by seven that evening. I was tired, but I was also looking forward to getting home, and so Richard and I took the minimum number of breaks and pit-stops and drove home as quickly as we could. During the journey back our trip was punctuated by several telephone calls from newspapers, news agencies and telephone interviews with various radio stations as far afield as Toronto and Barcelona.

In between interviews Richard and I discussed our new found fame. After years of trying, it seemed, that almost by accident our adventure had become the most talked about novelty news item in Britain's mass media - that week at least - and considering that we have no pretensions towards being anglers at all, we had become Britain's most famous fishermen. We had a gap of about three weeks before our next scheduled adventure - a trip up to Yorkshire with a team from UK Living (the cable TV Channel) in order to help the Yorkshire Constabulary investigate a

255

particularly horrible animal mutilation case, and it was clear that we had to do something to capitalise on this new level of media exposure.

It's ironic, I commented to Richard during the journey, that such a minor expedition as five days in Lancashire seemed to have captured the public attention far more than had our three week excursion to Puerto Rico in search of living vampires. However, beggars can't be choosers and this was obviously the publicity boost that we needed in order to establish a groundswell of support for the CFZ amongst the men, women and children in the street before launching the CFZ as a registered charity later in the year.

It was clear that we needed to do something - but what?

We were still pondering the matter when we arrived in Exeter just before seven that evening. Almost immediately we walked through the door we were pounced on by a Canadian radio station demanding a telephone interview, and so whilst Graham and John unpacked the two cars, Richard and I gave one of the most playful and jolly interviews of our media career. It did not, perhaps, reach the dizzy heights of silliness of the day before when Richard had done his best to convince an incredulous Sarah Burdon that with a name like Wisniewski, Pat was obviously a Russian Spy who, at the end of the Cold War had come to Lancashire to carry out secret genetic experiments on swans, whilst protecting his underwater laboratory by building a giant animatronic robot catfish which he used to scare away snoopers, but it came close.

By half past seven, however, I had kissed my daughter and was having an animated telephone conversation about our adventures with my father. Each day I had telephoned him on my mobile but partly due to pressure of work, and partly due to cost I had been forced to keep these daily bulletins short, functional and to the point. Now, at last we could talk properly and I think my father was even more surprised than us at the level of interest the story had gathered in the popular media. He had heard me talking on BBC Radio 4, on Classic FM, and had been given various pictures of us in local and national newspapers. He blustered about me needing a haircut, but I think that he was actually rather proud of our achievements, although totally bemused as to why his eldest son had suddenly become a minor celebrity just because he had spent five days camped out on the banks of a lake in Lancashire where someone had seen a big fish!

We laughed and joked, and then he went off to bed, and I had my dinner and logged onto the Internet to find Becky. She seemed overjoyed to see me and by the miracle of electronic communications we spent the night together in cyberspace lying in each other's virtual arms and caressing each other with words. The following morning, the sun was high in the sky when I finally drifted off to sleep, satiated and happy, secure in the knowledge that I wouldn't have to get up until I wanted to.

About three hours later I was rudely awakened by a telephone call from GMTV - the ITV breakfast show - who had decided in their infinite wisdom that they wanted us to go back to Southport, to spend the night in a hotel and to take part in a "Live from the Lakeside" Outside Broadcast the following morning.

This was the big break we had been wanting and Richard and I abandoned any pretence at having some time off and jumped on the first train to the north. During 1997-8 we had been regular guests on a BBC radio show, and during 2000/1 we had done a monthly two hour spot on Jeff Rense's syndicated US radio show. However, we had always wanted to become regular "house monster hunters" on one of the popular TV shows, and we felt that this could well be out chance.

We had appeared on the *Big Breakfast* on Channel 4, shortly after Richard had returned from Thailand in the autumn of 2000 but the show was in such disarray (it was cancelled not long after) that any hopes that we had nursed about becoming regulars with our own spot on the show were soon dispelled. However, fate had thrown us another chance and we were determined to do our damnedest to capitalise on it.

The journey to the north was so horrific that, if there hadn't been so much riding on it we would probably have given up. Just after we had passed Taunton on the first leg of the journey the train shuddered to a halt and a nasal voice came over the tannoy and proclaimed that "there had been a minor software problem" and that the train was going to be delayed. As there were clouds of smoke gushing out of the engine it seemed more likely that in the new British-Railspeak, "software problem" meant that the engine had burst into flames.

We were an hour late arriving in Bristol where as a result of the unforeseen conflagration in the engine we had to make an unscheduled change of trains. The replacement train was about half an hour late, and unbearably crowded. There was standing room only throughout most of the train and the whole scene was reminiscent of a jerky black and white newsreel of refugees crowded into a train escaping from eastern Europe in the years immediately following the last war.

However, even now the passengers were reasonably calm. They were British after all, and suffering in queues or execrable public transport is what the British do best. I had been lucky enough to secure a seat and was just settling back comfortably - half expecting to hear someone singing a Vera Lynn song as the "spirit of the Blitz" was so much in evidence - when another message came across the tannoy and this train too shuddered to a halt not two hundred yards outside Bristol Temple Meads station.

Apparently this engine too had broken down, and as the beleaguered train limped back into the station we were told that the train would have to be vacated before it could be turned around and another engine attached. By this time the assemblage of passengers was inclined strongly towards rebelliousness and there were angry murmuring amongst the populace as we all got off the train.

However before we had even all disembarked, the bloke on the tannoy had changed his mind and told us all to get back on the train again. This we did. By the time we had finally resumed our journey (at a speed roughly achievable by an arthritic snail suffering from a degenerative disease of the spleen), we were over two hours late. Half an hour outside Gloucester there was another software failure, all the toilets overflowed and the electronic doors of the carriages, and the toilets jammed either open or shut and remained like that for the remainder of the journey. We arrived in Birmingham two hours late in an overcrowded train which stank of urine and upon which (possibly wisely) no alcohol had been sold despite the fact that the bar was well stocked. "I wonder if Tim really *could* make the trains run on time?" I mused to Richard who just nodded back at me grimly.

We changed trains again and the rest of the journey was uneventful. A harassed looking bloke in a chauffeur's outfit met us in a limo at Manchester and drove us to the Sacrisbrick Hotel in Southport where we arrived just after midnight. Richard and I made for the bar where we charged several plates of sandwiches and some fairly hefty drinks to the GMTV bar tab and then sloped off to bed.

The next morning I was awoken by Graham at 5 am. He had been up all night waiting to see us on television and was fairly impressively drunk. I thanked him for his kindness in giving me a wakeup call and went back to sleep for twenty minutes before rising, showering, donning my suit and going down to the lobby to wait for the limo. Richard joined me just after 6 am. He was utterly exhausted and could hardly string two coherent words together. The same limo driver as the night before (looking remarkably fresh) was there to meet us and drove us the eight miles down the well-remembered roads towards Martin Mere. Unlike the idyllic weather that we had enjoyed during our sojourn at the lake, it was peeing it down as we drove into the WWT reserve car park and looked for the rest of the crew.

There was no-one to be seen.

In an appalling act of social vandalism for which I can never apologise profusely enough I was forced to telephone Pat's wife - the lovely Louise - on my mobile and ask her to open the back gates for us. She came down to the gate to meet us half asleep and blearily opened the gates to let the long black limo in. She looked mildly surprised to see us but welcomed us with a smile and then went back to bed. However there was no-one to be seen anywhere on the lake, and I began to worry.

Telephoning my contact at GMTV (it was his day off and he was in bed but tough), I got hold of the production manager's mobile and ten minutes later we were sitting in a large Satellite van at the other side of the lake drinking coffee with the production crew. We did a total of three light hearted RV broadcasts, and then made our way back to Southport. During the final segment Pat and Carl Lamb came out to meet us. Carl had been gutted that his previously arranged annual leave had been booked for the second half of our expedition and that he had left us on Friday night unaware of the outcome of our investigations. It was good to see the two of them again and we greeted each other warmly.

Apparently there had been a further spate of newspaper reports in the 48 hours which had elapsed since our departure, and whilst Carl with his Marketing Manager hat on was obviously overjoyed, we got the impression that Pat, though genuinely glad to see Richard and me again, was secretly wishing that the media furore would come to an end and that he could have his quiet little nature reserve back. What, I think, had particularly irked him was a piece in *The Sun* the previous day, quoting claims that "Marty" could be a man-eater.

I had felt much more at home at the lake when I had been there a few days before. Living and working arm in arm with my friends and colleagues. Working together to solve the mystery of the monster of the Mere. Now, although Richard was in his CFZ Uniform and as always was a cryptozoologist pure and simple, I was in a pinstriped suit and the explorer, and fortean zoologist had been replaced by a media pundit desperately trying to increase his media presence. It was not a role I was comfortable with although it was one that I knew, sadly, that I would have to embrace more and more in the future if the CFZ is to be the success it deserves to be.

We said our goodbyes for the second time in three days, and a taxi driven by an obliging little bloke called Harry who talked nineteen to the dozen, and in Douglas Adams's words, could not only have talked to hind leg off a donkey but could have persuaded him to go for a walk afterwards, drove us back to Southport. There we had breakfast with Lynda Matthews and the two girls, and I did a radio interview for BBC Three Counties radio, on my phone live from the breakfast restaurant at the Scarisbrick Hotel (during which I playfully did a vox pop interview with the waitresses), before renewing our relationship with Harry who drove us back to Manchester to catch the train back to Exeter.

During the long journey I drifted back to sleep, and in that hazy ur-space between full consciousness and sleep I looked back over the evidence that we had collected.

We had established a proper link between Frank Buckland and the town of Southport. We knew that he had visited the aquarium at the Winter Gardens. The two major engineering projects in the area at the time had been the Winter Gardens and the Leeds-Liverpool Canal, and it does not take too big a leap of faith to surmise that the wealthiest men of the neighbourhood (for example the Scarisbrick family) were amongst the shareholders of both projects. We had, after all, established that both projects had a large number of shareholders. We had been unable to establish that Sir Charles Scarisbrick was, indeed, a major shareholder of both attractions, but it seemed eminently likely. What if Scarisbrick and Buckland knew each other? Perhaps Buckland had stayed with Scarisbrick on the occasion that he had made the acquaintance of "Uncle Tom" the Alligator.

We did not know whether Scarisbrick (who was undoubtedly the major landowner in the Martin Mere region), had been a member of the Acclimatisation Society, but we can hazard a pretty informed guess that *some-one* in the area probably had been. Indeed, the very fact that Buckland had visited what was, at that time, a very insignificant seaside resort, suggests that he *had* indeed had acquaintances in the area. It is a known fact that Buckland was fond of travelling around the country visiting friends and involving them in his extravagant schemes.

Buckland's involvement in the introduction of wels to waters around the country was well known, and it would seem not at all unlikely that he had introduced specimens to waterways in west Lancashire. If he had done so, what better place than the new canal? Or possibly one of the deep pools in what had once been the largest lake in England.

Although I couldn't prove it, my personal hunch was that Buckland had indeed introduced the creatures to the Liverpool-Leeds Canal. Not only had that stretch of water been, as we have seen, financed by a heterogeneous cross section of the great and the good of British society (amongst whom must surely have been at least one of Buckland's cronies), but it was (as we have also seen) a remarkable piece of engineering stretching right across the Pennines. I felt sure that Buckland would have jumped at the chance of introducing a spectacular new species to a waterway which cut a literal swathe across much of northern England.

The historical record showed that the weather in west Lancashire had often produced flooding and that the canal had burst its banks, (as indeed had the other waterways in the area), with monotonous regularity. What if, one of those occasions during the last thirty years, a large and elderly fish - which could well have been one of Buckland's original importation – had, in the words of the song my Dad used to sing to me as a small boy "swam and swam right over the dam", and eventually ended up in the lake at the WWT Reserve.

There was no doubt that the interlocking series of waterways across what had once been the bottom of the great lake, would make this migration eminently possible.

Although there was no way that we could prove it, I felt fairly certain that we had solved the case. "Marty" was indeed quite possibly the largest, and the oldest, freshwater fish in England. I grinned cheerfully to

261

myself and went back to sleep.

Arriving at Manchester Station I bought a copy of *The Guardian* . Inside, to my fury I read the following jokey and unfunny piece:

'*Real name:*
Wels catfish or Silurus glanis.

Appearance:
Ghastly, oily, leering creature from the deep, sporting hideous piggy little eyes and a number of curious, but undoubtedly malevolent, protuberances from its funny little forehead.

Temperament:
Evil. Really, really evil.

Blimey, how do you know?
Just look at it... that hideous, gaping mouth, the evil glint of its swivelling eyes...

It just looks like a fish to me.
That's what Jonah said, my friend. This boy is baad. He's mean. He's chomping swans for breakfast without so much as a by-your-leave from Her Maj. And now he's coming after your children.

What on earth are you talking about?
OK - simple words only. This is the Monster of Martin Mere. He lives in a 15ft-deep lake in Ormskirk, Lancs, and has done for more than 100 years...

How do they know?
They guessed, of course. Anyway, during those 100 years, while the anglers of Lancashire were shamefully neglecting their angling duties, the Monster has grown to a staggering 8ft long.

Average size for a monster, I should have thought.
Oh, shut up. Now, out of a sense of hubris, hunger or just plain boredom, he's started mauling swans
.
That's not very nice.
Monsters aren't. Anyway, now scientists are trying to catch him before he turns his attentions to homo sapiens. Apparently Wels catfish have been known to kill people.

With a stare?
Um, a well-timed gobble, presumably.
So it's Monster steaks all round!
Don't be silly, that would be cruel. If they ever do get him (and why break the habit of a century now?) Monster will be rehomed in an aquarium. A big one. With no swans.

262

<u>*Do say:*</u>
"Honestly, it was thiiis biiig..."

<u>*Don't say:*</u>
"I've heard you might eat small children. Do you take bookings?"

<u>*Not to be confused with:*</u>
I think we'd better not go there, don't you?

"You know", I thought to myself, "The journalists who pander to the tastes of the chattering classes are no better than the folk who dole out drivel in the tabloids." I thought sadly about these people. I had met many over the years. They live in London, they go from wine bar to night club living a hectic and frenetic life with very little substance to it. It is easy for them to mock our efforts, and possibly after the story in the previous day's *Sun* they felt that they needed mocking. However, it had been us that had put together a £2,500 pound expedition without any outside finance. It had been us who had gathered together a collection of specialist equipment and learned how to use it, and furthermore it had been us who had seen the elusive "Monster of the Mere".

Not for them the joy of spending a night asleep in the front seat of an elderly Jaguar and waking to the sight of a young stoat cavorting across the path only a few feet away.

Not for them the adventure of chasing a fearsome monster through the murky waters of the Mere.

Not for them the camaraderie of sitting beneath the stars waiting for a giant fish to surface on the moonlit lake.

I bet they had never bought life jackets from an armed airman on a secret airforce base, and I would wager a couple of cases of scotch that none of them had ever bought a rubber dinghy from a drug dealer in the middle of a sheltered housing project for the mentally subnormal!

I looked out of the window as the train sped through the grey splurge of the rainy Midlands countryside. People like that may have the money. People like that may have the job security. I may be a middle aged manic depressive. I may have just lost my mother. I may have serious heart disease. I may be in an unresolveable conflict with my best friend, and be emotionally involved with a woman I have never met on the wrong side of the Atlantic. I may never have the money that a *Guardian* journalist living in a posh London flat will have, and I may end my days living in Bohemian squalour with my daughter and a houseful of reptiles, but you know what? I would not swap my life with anyone.

If it comes to a choice between me, and a overpaid yuppie writing snide nonsense in a London newspaper office, I know which of us is the lucky one!

******

Deep in the lake the great fish stirred from its resting place. Slowly it swam from where it had been hiding for the past few days and made its way back towards the shallow water where it had made its home for the previous few years. The people who had disturbed its home for the last week had gone and its life could continue in much the same way as it had for over a hundred years. Soon it would be winter and the swans would be back…

- FIN -

- APPENDIX ONE -
THE MERMAID OF
MARTIN MEER

The following extract is taken from the second volume of *Traditions of Lancashire* by John Roby, (George Routledge and Sons, London, 1882)

Now the dancing sunbeams play
()' er the green and glassy sea
Come with me, and we will go
Where the rocks of coral grow.

LITTLE needs to be said by way of introduction or explanation of the following tale. Martin Meer is now in the process of cultivation; the plough and the harrow leave more enduring furrows on its bosom. It is a fact, curious enough in connection with our story, that some years ago, in digging and draining, a canoe was found here. How far this may confirm our tradition, we leave the reader to determine. It is scarcely two miles from Southport; and the botanist, as well as the entomologist, would find themselves amply repaid by a visit.

MARTIN MEER, the scene of the following story, we have described in our first series of *Traditions,* where Sir Tarquin, a carnivorous giant, is slain by Sir Lancelot of the Lake. These circumstances, and more of the like purport on this subject, we therefore omit, as being too trite and familiar to bear repetition. We do not suppose the reader to be quite so familiar with the names and fortunes of Captain Harrington and Sir Ralph Molyneux, though they had the good fortune to be born eleven hundred years later, and to have seen the world, in consequence, eleven hundred years older - we wish we could say wiser and better-tempered, less selfish and less disposed to return hard knocks, and to be corrupted with evil communications. But man is the same in all ages. The external habits and usages of society change his mode of action - clothe the person and passions in a different garb; but their form and substance, like the frame they inhabit, are unchanged, and will continue until this great mass of intelligence, this mischievous compound of good and evil, this round rolling earth, shall cease to swing through time and space - a mighty pendulum, whose last stroke shall announce the end of time, the beginning of eternity!

Our story gets on indifferently the while; but a willing steed is none the worse for halting. Harrington and his friend Sir Ralph were spruce and well-caparisoned cavaliers, living often about court towards the latter end of Charles the Second's reign. What should now require their presence in these extreme regions of the earth, far from society and civilisation, it is not our business to inquire. It sufficeth for our story that they were here, mounted, and proceeding at a shuffling trot along the flat, bare, sandy region we have described.

How sweetly and silently that round sun sinks into the water !" said Harrington.

"But doubtless," returned his companion, "if be were fire, as thou sayest, the liquid would not bear his approach so meekly; why, it would boil if he were but chin-deep in yon great seething-pot."

"Thou art quicker at a jest than a moral, Molyneux," said the other and graver personage; "thou canst not even let the elements escape thy gibes. I marvel how far we are from our cousin Ireland's at Lydiate. My fears mislead me, or we have missed our way. This flat bosom of desolation hath no vantage-ground whence we may discern our path; and we have been winding 'about this interminable lake these two hours."

"Without so much as a blade of grass or a tree to say 'Good neighbour' to," said Molyneux, interrupting his companion's audible reverie.

"Crows and horses must fare sumptuously in these parts."

"This lake, I verily think, follows us; or we are stuck to its side like a lady's bauble."

"And no living thing to say 'Good-bye,' were it fish or woman."

"Or Mermaid, which is both." Scarcely were the words uttered when Harrington pointed to the water.

"Something dark comes upon that burning track left on the surface by the sun's chariot wheels."

"A fishmonger's skiff belike," said Sir Ralph.

They plunged through the deep sandy drifts towards the brink, hastening to greet the first appearance of life which they had found in this region of solitude, At a distance they saw a female floating securely, and apparently without effort, upon the rippling current. Her form was raised half-way above the water, and her long hair hung far below her shoulders. This she threw back at times from her forehead, smoothing it down with great dexterity. She seemed to glide on slowly, and without support; yet the distance prevented any very minute observation,

"A bold swimmer, 0' my troth I" said Molyneux ; "her body tapers to a fish's tail, no doubt, or my senses have lost their use."

Harrington was silent, looking thoughtful and mysterious.

"I'll speak to yon sea-wench."

"For mercy's sake, hold thy tongue. If as I suspect-and there be such things, 'tis said, in God's creation-thou wilt "

But the tongue of this errant knight would not be stayed; and his loud musical voice swept over the waters, evidently attracting her notice, and for the first time. She threw back her dark hair, gazing on them for a moment, when she suddenly disappeared. Harrington was sure she bad sunk; but a jutting peninsula of sand was near enough to have deceived him, especially through the twilight, which now drew on rapidly.

267

"And thou bast spoken to her!" said he gravely "then be the answer thine I"

"A woman's answer were easier parried than a swordthrust, methinks; and that I have hitherto escaped."

"Let us be gone speedily. I like not yon angry star spying out our path through these wilds."

"Thou didst use to laugh at my superstitions; but thine own, I guess, are too chary to be meddled with."

"Laugh at me an' thou wilt," said Harrington : "when Master Lilly cast my horoscope he bade me ever to eschew travel when Mars comes to his southing, conjunct with the Pleiades, at midnight-the hour of my birth. Last night, as I looked out from where I lay at Preston, methought the red warrior shot his spear athwart their soft scintillating light and as I gazed, his ray seemed to ride half-way across the heavens. Again he is rising yonder."

"And his meridian will happen at midnight?"

"Even so," replied Harrington.

"Then gallop on. I'd rather make my supper with the fair dames at Lydiate than in a mermaid's hall."

But their progress was a work of no slight difficulty, and even danger. Occasionally plunging to the knees in a deep bog, then wading to the girth in a hillock of sand and prickly bent grass (the *Arundo arenana,* so plentiful on these coasts), the horses were scarcely able to keep their footing-yet were they still urged on. Every step was expected to bring them within sight of some habitation.

"What is yonder glimmer to the left?" said Molyneux. "If it be that hideous water again, it is verily pursuing us. I think I shall be afraid of water as long as I live."

"As sure as Mabomet was a liar, and the Pope has excommunicated him from Paradise, 'tis the same still, torpid, dead-like sea we ought to have long since passed."

" Then have our demonstrations been in a circle, in place of a right line, and we are fairly on our way back again.

Sure enough there was the same broad, still surface of the Meer, though on the contrary side, mocking day's last glimmer in the west. The bewildered travellers came to a full pause. They took counsel together while they rested their beasts and their spur-rowels; but the result was by no means satisfactory. One by one came out the glorious throng above them, until the heavens grew light with living hosts, and the stars seemed to pierce the sight, so vivid was their brightness.

"Yonder is a light, thank Heaven!" cried Harrington.

"And it is approaching, thank your stars!" said his companion. " I durst not stir to meet it, through these perilous paths, if our night's lodging depended on it."

The bearer of this welcome discovery was a kind-hearted fisherman, who carried a blazing splinter of antediluvian firewood dug from the neighbouring bog; a useful substitute for more expensive materials.

It appeared they were at a considerable distance from the right path, or indeed from any path that could be travelled with safety, except by daylight. He invited them to a lodging in a lone hut on the borders of the lake, where he and his wife subsisted by eel-catching and other precarious pursuits. The simplicity and openness of his manner disarmed suspicion. The offer was accepted, and the benighted heroes found themselves breathing fish-odours and turf-smoke for the night, under a shed of the humblest construction. His family consisted of a wife and one child only; but the strangers preferred a bed by the turf-embers to the couch that was kindly offered them.

The cabin was built. of the most simple and homely materials. The walls were pebble-stones from the sea-beach, cemented with clay. The roof-tree was the wreck of some unfortunate vessel stranded on the coast. The whole was thatched with star-grass or sea-reed, blackened with smoke and moisture.

"You are but scantily peopled hereabouts," said Harrington, for lack of other converse.

269

"Why, ay," returned the peasant; "but it matters nought our living is mostly on the water."

"And it might be with more chance of company than on shore; we saw a woman swimming or diving there not long ago."

"Have you seen her?" inquired both man and dame with great alacrity.

"Seen whom !" returned the guest.

"The Meer-woman, as we call her."

"We saw a being, but of what nature we are ignorant, float and disappear as suddenly as though she were an inhabitant of you world of waters."

"Thank mercy! Then she will be here anon."

Curiosity was roused, though it failed in procuring the desired intelligence. She might be half-woman half-fish for aught they knew. She always' came from the water, and was very kind to them and the babe. Such was the sum of the information;, yet when they spoke of the child there was evidently a sort of mystery and alarm, calculated to awaken suspicion.

Harrington looked on the infant. It was on the woman's lap asleep, smiling as it lay; and an image of more perfect loveliness and repose he had never beheld It might be about a twelvemonth old; but its dress did not correspond with the squalid poverty by which it was surrounded.

"Surely this poor innocent has not been stolen ," thought he. The child threw its little hands towards him as it awoke; and he could have wept. Its short feeble wail had smitten him to the heart.

Suddenly they heard a low murmuring noise at the window.

"She is there," said the woman ; "but she likes not the presence of strangers. Get thee out to her, Martin, and persuade her td come in."

The man was absent for a short time. When he entered, his face displayed as much astonishment as it was possible to cram into a countenance so vacant.

She says our lives were just now in danger, and that the child's enemies are again in search; but she has put them on the wrong scent. We must not tarry here any longer; we must remove, and that speedily. But she would fain be told what is your business in these parts, if ye. are so disposed"

"Why truly," said Harrington, "our names and occupation need little secrecy. We are idlers at present, and having kindred in the neighbourhood, are on our way to the Irelands at Lydiate, as we before told thee. Verily, there is but little of either favour or profit to be had about court now-a-days. Nought better than to loiter in hall and bower, and fling our swords in a lady's lap. But why does the woman ask? Hath he some warning to us? or is there already a spy upon our track?"

"I know not," said Martin ; "but she seems mightily afeard of the child."

If she will entrust the babe to our care," said Harrington, after. a long pause, " I will protect it.

The shield of the Harringtons shall be its safeguard."

The fisherman went out with this message; and on his return it was agreed that, as greater safety would be the result, the child should immediately be given to Harrington. A solemn pledge was required by the unseen visitant that the trust should be surrendered whenever, and by whomsoever demanded; likewise a vow of inviolable secrecy was exacted from the parties that were present. Harrington drew a signet from his finger; whoever returned it was to receive back the child. He saw not the mysterious being to whom it was sent; but the idea of the Meerwoman, the lake, and the untold mysteries beneath its quiet bosom, came vividly and painfully on his recollection.

Long after she had departed, the strange events of the evening kept them awake. Inquiries were now answered without hesitation. Harrington learned that the "Meer-woman's" first appearance was on a cold wintry day, a few months before. She did not crave protection from the dwellers in the hut, but seemed rather to command it. Leaving the infant with them, and promising to return shortly, she seemed to vanish upon the lake, or rather, she seemed to glide away on its surface so swiftly that she soon disappeared. Since then she had visited them thrice, supplying them with a little money and other necessaries ; but they durst not ques-

271

tion her, she looked so strange and forbidding.

In the morning they were conducted to Lydiate by the fisher-man, who also carried the babe. Here they told a pitiable story of their having found the infant exposed, the evening before, by some unfeeling mother; and, strange to say, the truth was never divulged until the time arrived when Harrington should render up his trust.

Years passed on. Harrington saw the pretty foundling expand through every successive stage from infancy to childhood growing lovelier as each year unfolded some hidden grace, and the bloom brightened as it grew. He had married in the interval, but was yet childless. His lady was passionately fond of her charge, and Grace Harrington was the pet and darling of the family. No wonder their love to the little stranger was growing deeper, and was gradually acquiring a stronger hold on their affections. But Harrington remembered his vow: it haunted him like a spectre. It seemed as though written with a sunbeam on his memory; but the finger of death pointed to its accomplishment. It will not be fulfilled without blood, was the foreboding that assailed him. His lady knew not of his grief, ignorant happily of its existence, and of its source.

Their mansion stood on a rising-ground but a few miles distant from the lake. He thus seemed to hover instinctively on its precincts; though, in observance of his vow, he refrained from visiting that lonely hut, or inquiring about its inhabitants. Its broad smooth bosom was ever in his sight; and when the sun went down upon its wide brim his emotion was difficult to conceal.

One soft, clear evening, he sat enjoying the calm atmosphere with his lady and their child. The sun was nigh setting, and the lake glowed like molten fire at his approach.

'Tis said a mermaid haunts yon water," said Mrs. Harrington; "I have heard many marvellous tales of her a few years ago. Strange enough, last night I dreamed she took away our little girl, and plunged with her into the water. But she never returned."

How I should like to see a mermaid!" said the playful girl. "Nurse says they are beautiful ladies with long hair and green eyes. But...." and she looked beseechingly towards them .."we are always forbidden to ramble towards the Meer."

Harrington, the night wind makes you shiver. You are ill ! "

"No, my love. But this cold air comes wondrous keen across my bosom;" said he, looking wistfully on the child, who, scarcely knowing why, threw her little arms. about his neck, and wept.

"My dream, I fear, hath strange omens in it," said the lady thoughtfully.

The same red star shot fiercely up from the dusky horizon; the same bright beam was on the wave; and the mysterious incidents of the fisher-man's hut came like a track of fire across Harrington's memory:

"Yonder is that strange woman again that has troubled us about the house these three days," said Mrs. Harrington, looking out from the bal-cony; ".we forbade her yesterday. . She comes hither with no good in-tent."

Harrington looked over the balustrade. A female stood beside a pillar, gazing intently towards him. Her eye caught his own; it was as if a basi-lisk had smitten him. Trembling, yet fascinated:', he could not turn away his glance; a smile passed on her dark red visage - a grin of joy at the discovery.

"Surely," thought he, "'tis not the being who claims my child! " But the woman drew something from her hand, which, at that distance, Harring-ton recognised as his pledge; His lady saw not the signal; without speak-ing, he obeyed.

Hastening down-stairs, a private audience confirmed her demand, which the miserable Harrington durst not refuse.

Two days he was mostly in private. Business with the steward was the ostensible motive. He, had sent an urgent message to his friend Moly-neux, who, on the third day, arrived at H___ where they spent many hours in close consultation. The following morning Grace came running in after breakfast. She flung her arms about his neck, "Let me not leave you today," she sobbed aloud.

"Why, my love ?" said Harrington, strangely disturbed at the request.

"I do not know" replied the child, pouting.

"Today I ride out with Sir Ralph to the Meer, and as thou hast often wished - because it was forbidden, I guess - thou shalt ride with us a short distance; I will toss thee on before me, and away we'll gallop - like the Prince of Trebizond on the fairy horse."

"And shall we see the mermaid?" said the little maiden quickly, as though her mind had been running on the subject.

"I wish the old nurse would not put such foolery in the girl's head," said Mrs. Harrington impatiently. "There be no Mermaids now, my love."

"What! not the mermaid of Martin Meer?" inquired the child, seemingly disappointed.

Harrington left the room, promising to return shortly.

The morning was dull, but the afternoon broke out calm and bright. Grace was all impatience for the ride; and Rosalind, the favourite mare, looked more beautiful than ever in her eyes. She bounded down the terrace at the first sound of the horses' jeet, leaving Mrs. Harrington to follow.

The cavaliers were already mounted, but the child suddenly drew hack.

"Come, my love," said Harrington, stretching out his hand; "look how your pretty Rosalind bends her neck to receive you.

Seeing her terror, Mrs. Harrington soothed these apprehensions, and fear was soon forgotten amid the pleasures she anticipated.

"You are back by sunset, Harrington?"

Fear not, I shall return," replied he; and away sprang the pawing beasts down the avenue. The lady lingered until they were out of sight. Some unaccountable oppression weighed down her spirits; she sought her chamber, and a heavy sob threw open the channel which hitherto had restrained her tears. They took the nearest path towards the Meer, losing sight of it as they advanced into the low flat sands, scarcely above its level, When again it opened into view, its wide waveless surface lay before them, reposing in all the sublimity of loneliness and silence, The rapture of the child was excessive. She surveyed with delight its broad unruffled bosom, giving back the brightness and glory of that heaven to which it looked; to her it seemed another sky and another world, pure and spotless as the imagination that created it.

They entered the fisherman's hut; but it was deserted. Years had probably elapsed since the last occupation. Half burnt turf and bog-wood lay on the hearth, but the walls were crumbling down with damp and decay.

The two friends were evidently disappointed. At times they looked out anxiously, but in vain, as it might seem; for they again sat down, silent and depressed, upon a turf-heap by the window, while the child ran playing and gambolling towards the beach.

Harrington sat with his back to the window, when suddenly the low murmuring noise he had heard on his former visit was repeated. He turned pale.

"Thou art not alone; and where is the child?" or words to this purport, were uttered in a whisper. He started aside; the sound, as he thought, was close to his ear. Molyneux heard it too.

"Shall I depart?" said he, cautiously; "I will take care to keep within call,"

"Nay," said his friend, whispering in his ear, "thou must ride out of sight and sound too, I am afraid, or we shall not accomplish our plans for the child's safety. Depart with the attendants; I fear not the woman. Say to my lady I will return anon."

With some reluctance Sir Ralph went his way homeward; and Harrington was left to accomplish these designs without assistance.

Immediately he walked out towards the shore; but he saw nothing 6f the child, and his heart misgave him, He called her; but the sound died with its own echo upon the waters. The timid rabbit fled to its burrow, and the sea-gull rose from her gorge, screaming away heavily to her mate; but the voice of his child returned no more.

Almost driven to frenzy, he ran along the margin of the lake to a considerable distance, returning after a fruitless search to the hut, where he threw himself on the ground. In the agony of his spirit he lay with his face to the earth, as if to hide his anguish as he wept.

How long he remained was a matter of uncertainty. On a sudden, instantaneously with the rush that aroused him, he felt his arms pinioned, and that by no timid or feeble hand. At the same moment a bandage was thrown over his eyes, and he found himself borne away swiftly into a boat. He listened for some time to the rapid stroke of the oars. Not a word was spoken from which he could ascertain the meaning of this outrage. To his questions no reply was vouchsafed, and in the end he forbore inquiry-the mind wearied into apathy by excitement and its consequent exhaustion.

The boat again touched the shore, and he was carried out. The roar of the sea had for some time been rapidly growing louder as they neared the land. He was now borne along over hillocks of loose sand to the sea-beach, when he felt himself fairly launched upon the high seas. He heard the whistling of the cordage; the wide sail flap to the wind with the groan of the blast as it rushed into the swelling canvas ; then he felt the billows prancing under him, and the foam and spray from their huge necks as they swept by. It was not long ere he heard the sails lowered; and presently they were brought up alongside a vessel of no ordinary bulk. Harrington was conducted with little ceremony into the cabin; the bandage was removed from his eyes, and he found himself in the presence of a weather beaten tar, who was sitting by a table, on which lay a cutlass and a pair of richly-embossed pistols..

"We have had a long tug to bring thee to," said the captain; "but we always grapple with the enemy in the long-run. If thou hast aught to say why sentence of death should not pass on thee - ay, and be executed straightway too - say on. What I not a shot in thy locker? Then may all such land-sharks perish, say I, as thus I signify thy doom." He examined his pistols with great nicety as he spoke. Harrington was dumb with amazement, whilst his enemy surveyed him with a desperate and determined glance. - At length he stammered forth - "I am ignorant of thy meaning; much less can I shape my defence. Who art thou?"

The other replied, in a daring and reckless tone - "I am the Free Rover, of whom thou hast doubtless heard.

My good vessel and her gallant crew ne'er slackened a sky-raker in the chase, nor hacked a mainsail astern of the enemy. But pirate as I am-hunted and driven forth like the prowling wolf, without the common - rights and usages of my fellow-men - I have yet their feelings. I had a child

Thy fell, unpitying purpose, remorseless monster, hath made me child-less I But thou hast robbed the lioness of her whelp, and thou art in her gripe!"

"As my hope is to escape thy fangs, I am innocent of the crime."

"Maybe thou knowest not the mischief thou hast inflicted; but thy guilt and my bereavement are not the less. My child was ailing; we were off this coast, when we sent her ashore secretly until our return. A fisher-man and his wife, to whom our messenger entrusted the babe, were driven forth by thee one bitter night without a shelter. The (child per-ished; and its mother chides my tardy revenge."

"'Tis a falsehood I" cried Harrington, "told to cover some mischievous design. The child, if it be thine, was given to my care - by whom I know not. I have nurtured her kindly; not three hours ago, as I take it, she was in yonder hut; but she has been decoyed from me; and I am here thy pris-oner, and without the means of clearing myself from this false and mali-cious charge."

The captain smiled incredulously.

"Thou art lord of yonder soil, I own; but thou shouldest have listened to the cry of the helpless. I have here a witness who will prove thy story false-the messenger herself. Call hither Oneida$_1$" said he, speaking to the attendants. But this personage could not be found.

"S he has gone ashore in her canoe,'" said the pirate; "and the men never question her. She will return mid-watch. Prepare thou showedst no mercy, and I have sworn.

Harrington was hurried to a little square apartment; which an iron grating sufficiently indicated to be the state prison.

The vessel lay at anchor; the intricate soundings on that dangerous coast rendered her perfectly safe from attack, even if she had been discovered.

He watched the stars rising out, calm and silently, from the deep "Ere yon glorious orb is on the zenith," thought he, "I may be-what?" He shrank from the conclusion. "Surely the wretch will not dare to execute his audacious threat?" He again caught that red and angry star gleaming portentously on him. It seemed to be his evil genius; its malignant eye appeared to follow out his track, to haunt him, and to beset his path continually with suffering. and danger. He stood by the narrow grating, feverish and apprehensive; again he heard that low murmuring voice which he too painfully recognised. The mysterious being of the lake stood before him.-

"White man "-she spoke in a strange and uncouth accent

" the tree bows to the wing of the tempest-the roots look upward the wind sighs past its withered trunk-the song of the warbler is heard no more from its branches, and the place of its habitation is desolate. Thine enemies have prevailed. I did it not to compass thine hurt : I knew not till now thou wert in their power; and I cannot prevent the sacrifice."

"Restore the child, and I am safe," said Harrington, trembling in his soul's agony at every point; "or withdraw thy false, thine accursed accusations."

Thou knowest not my wrongs and my revenge ! Thou seest the arrow, but not the poison that is upon. it. The maiden, whose race numbers a thousand warriors, returns not to her father's tribe ere she wring out the heart's life-blood from her destroyer. Death were happiness to the torments we inflict on him and the woman who hath supplanted me. And yet they think Oneida loves them-bends like the bulrush when the wind blows upon her, and rises only when he departs. What I give back the child? She hath but taken my husband and my bed; as soon might ye tear the prey from the starved hunter. This night will I remove their child from them to depart, when a few moons are gone, it -. may be to dwell again with my tribe in the wigwam and the forest."

"But I have not wronged thee."

"Thou art of their detested race Yet would I not kill thee."

"Help me to escape."

"Escape" said this untamed savae, with a laugh which went with a shudder to his heart. "As soon might the deer dart from the hunter's rifle as thou from the cruel pirate who hast pronounced thy death.I could tell thee such deeds of him and. these bloody men as would freeze thy bosom, though it were wide and deep. as the lakes of my country. Yet I loved him once He came a prisoner to my father's hut. I have spilled my best blood for his escape. I have borne him where the white man's feet never trod - through forests, where aught but the Indian or the wild beast would have perished. I left my country and my kin-the graves of my fathers-and how hath he requited me? He gave the ring of peace to the red woman; but when he saw another and a fairer one of thy race, she became his wife; and from that hour Oneida's love was hate and I have waited and not complained, for my revenge was sure! And shall I now bind the healing leaf upon the wound? Draw the arrow from the flesh of mine enemies? Thou must die I for my revenge is sweet."

"I will denounce thee to him, fiend I will reveal"

"He will not believe thee. His eye and ear are sealed. He would stake his life on my fidelity. He knows not of the change."

"But he will discover it monster, when thou art gone He will track thee to the verge of this green earth and the salt sea, and thou shalt not escape."

With a yell of unutterable scorn, she cried - "He may track the wild bee to its nest, and the eagle to his eyrie, but he discerns not one footprint of Oneida's path

The pangs of death seemed to be upon him. He read his doom in,the kindling eye and almost demoniac looks of the being who addressed him. She seemed like some attendant demon waiting to receive his spirit. His brain grew dizzy. Death would have been welcome in comparison with the horrors of its anticipation. He would have caught her; but she glided from his grasp, and he was again left in that den of loneliness and misery. How long he knew not; his first returning recollection was the sound of bolts and the rude voices of' his jailers.

In this extremity the remembrance of that Being in whom, and from whom, are all power arid mercy, flashed on his brain like a burst of hope-like a sunbeam on the dark ocean of despair.

279

"God of my fathers, hear !" escaped from his lips in that appalling moment. His soul was calmed by the appeal. Vain was the help of man, but he felt as if supported and surrounded by the arm of Omnipotence, while silently, and with a firm step, he followed his conductors.

One dim light only was burning above. Some half-dozen of the crew stood armed on the quarter-deck behind their chief; their hard, forbidding faces looked without emotion upon this
scene of unpitying, deliberate murder

To some questions from the pirate, Harrington replied by accusing the Indian woman of treachery.

"As soon yonder star, which at midnight marks our meridian, would prove untrue in its course."
Harrington shuddered at this ominous reference.

"I cannot prove mine innocence," said he; "but I take yon orb to witness that I never wronged you or yours. The child is in her keeping."

"Call her hither, if she be returned," said the captain, "and see if he dare repeat this in her presence. He thinks to haul in our canvas until the enemy are under weigh, and then, Yoh, ho, boys, for the rescue. But we shall be dancing over the bright Solway ere the morning watch, and thy carcase in the devil's locker"

"If not for mine, for your own safety !"

"My safety and what care I though ten thousand teeth were grinning at me, through as many port-holes? My will alone bounds my power. Who shall question my sentence, which is death?"

He gnashed his teeth as he went on. "And your balls shall be too hot to hold your well-fed drones. Thy hearth, proud man, shall be desolate. I'll lay waste thy domain. Thy race, root and branch, will I extinguish; for thou hast made me childless."

The messenger returned with the intelligence that Oneida was not in the ship.
"On shore again, the ***** ! If I were to bind her with the main-chains, and an anchor at each leg, she would escape me to go ashore. No heed;

we will just settle the affair without her, and he shall drop quietly into a grave ready made, and older than Adam. I would we had some more of his kin; they should swing from the bowsprit, like sharks and porpoises, who devour even when they have had enough, and waste what they can't devour."

"Thou wilt not murder me thus, defenceless, and in cold blood."

"My child was more helpless, and had not injured Ye give no quarter to the prowling beast, and yet, like me, he only robs and murders to pre-serve his life. How far is it from mid-night?"

"Five minutes, and yon star comes to its southing," said the person he addressed.

"Then prepare; that moment marks thy death I"

The men looked significantly towards their rifles.

"Nay," cried this bloodthirsty freebooter, "my arm alone shall avenge my child."

He drew a pistol from his belt.

"Yonder is Oneida," sang out the man at the maintop; "she is within a cable's length."

"Heed her not. When the bell strikes, I have sworn thou shalt die !"

A pause ensued - a few brief moments in the lapse of time, but an age in the records of thought. Not a breath relieved the horror and intensity of that silence. The plash of a light oar was heard as a boat touched the vessel. The bell struck.

"Once " shouted the fierce mariner, and he raised his pistol with the sharp click of preparation.
"Twice "

The bell boomed again.
"Thrice"

281

"Hold" cried a female, rushing between the executioner and the con-demned. But the warning was too late;- the ball had sped, though not to its mark Oneida was, the victim

She fell, with a faint scream, bleeding on the deck. But Harrington was close locked in the arms of his little Grace. She had flown to him for protection, sobbing with joy.

The pirate seemed horror - struck at the deed. He raised Oneida, unloos-ing his neckcloth to staunch the wound

"The Great Spirit calls me :" she spoke with great exertion : "the green woods, the streams, land of my forefathers. Oh! I come." She raised her-self suddenly with great energy, looking towards Warrington, who yet knelt, guarded and pinioned-the child still clinging to him.

"White man, I have wronged thee, and I am the sacrifice. Murderer, be-hold thy child! She raised her eyes suddenly towards the pirate, who shook his head, supposing that her senses grew confused.

"It was for thy rescue", again she addressed Harrington. "The Great Spirit appeared to me: he bade me restore what I had taken away, and I should be with the warriors and the chiefs who have died in battle.

They hunt in forests from which the red-deer flies not, and fish in rivers that are never dry. But my bones shall not rest with my fathers I come. Lake of the woods, farewell!

She threw one look of reproach on her destroyer, and the spirit of Oneida had departed.

The pirate stood speechless and bewildered. He looked on the child - a ray of recollection seemed to pass over his visage.

Its expression was softened; and this man of outlawry and blood became gentle. The savage grew tame. The common sympathies of his nature, so long dried up, burst forth, and the wide deep flood of feeling and af-fection rolled on with it like a torrent, gathering strength by its own accu-mulation.

Years after, in a secluded cottage by the mansion of the Harringtons, dwelt an old man and his daughter. She soothed the declining hours of his sojourn. His errors and his crimes - and they were many and aggravated - were not unrepented of. She watched his last breath, and the richest lady of that land was THE PIRATE'S DAUGHTER.

- APPENDIX TWO -

MARTIN MERE AND ITS HISTORY

by the Rev W.T. Bulpit

(Originally published in the 'Southport Visitor' December 22nd 1906)

Martin Mere is often spoken of, yet few can locate it. In old days when the water was five miles long and two and a half miles wide it was easily recognised. This broad, spreading sheet of water has been drained and the hollow is occupied by grazing and arable land. The L and Y Railway skirts along it between Blowick and New Lane and travellers looking across the Mere may notice an old windmill.. This is situated about the centre of the Mere and now so lonely, carts have flocked to it and waited all night to get their batches ground. The Mere is skirted on its north side by the road from Crossens to Rufford, and from the high land at Holmes Wood a fine view across the Mere is obtained. The eastern termination to the lake is at Rufford near to the River Douglas, with which it was once connected by a short channel crossed by a bridge. The western boundary is hard to define. It varied with the season and rainfall but may be said to have washed the knoll on which Crossens stands. When very full of water I think outlets were to be found, both at Crossens and Hundred End, as well as into the Douglas.

The formation of the Mere points to times both of elevation and subsidence. I think that the original line of coast was further inland than at the present, and it was formed by the new red sandstone such as crops out at Halsall and Ormskirk. Upon this sandstone there formed a deposit of upper marl which forms the present floor of the deposits of the district. The glacial period brought a further deposit of clay intermingled with boulders. The vegetation which grew upon this clay formed a peat bed. A period of subsidence caused a blue clay (locally called Scotch), to settle upon this, and then the land being once more elevated, a forest grew upon the clay. This forest covered a wide area, for its roots, embedded in the clay, are found from the Ribble to the Dee, and are even submarine. Blown sand impeded the drainage of the area. Water accumulated and killed the trees. Then a south west gale snapped off the tree trunks, and swampy plants grew around them, and so an exceedingly thick bed of peat was formed. The sea washed up sand and silt, and thus a raised bank of ground separated the peat from the sea. Tree trunks are found in the peat, and also about 3 ft from its surface a layer remains of beech and hazel. The water of the lake reposed on the peat, and could not anywhere have been of great depth. Fish abounded, and so did water birds.

The drainage of the Mere, so far as we have it recorded, was commissioned by Mr Thomas Fleetwood of Bank Hall in 1692. He owned land at Mere Brow, and got the other landowners to make over their rights to him for a term of years. His chief work was cutting the Crossens Sluice. It is a mile and a half long and 24 feet wide. In 1693 as many as 2,000 men were engaged in the dry season. Success attended the effort, but blown sand soon obscured the outlet and fresh exertions were needed. Mr Fleetwood died in 1717 and chose to be buried at Churchtown where a memorial tablet declares "He made into solid land the immense Martinesian Marsh." In 1750 his lease expired and the landowners, not being careful, had their floodgates washed away in 1755, and though they were replaced and improved, through carelessness in management allowed the sluice to silt up, and then the Mere was for a long time partially flooded.

Mr Thomas Eccleston of Scarisbrick Hall leased what he could of the Mere and in 1781 recommence the drainage work. He spent two years in executing his works, and with the aid of flushing gates seemed so suc-

285

cessful that the Society of Arts presented him with its gold medal for his spirited effort. Danger now came from the eastern side of the Mere where the flooded Douglas and the Leeds-Liverpool Canal burst their banks and overwhelmed the meerland.

Mr Thomas Scarisbrick who succeeded his father Mr Eccleston, in 1809, did brave battle with the water, and when his sea gates, after standing 30 years, were swept away in 1819, renewed them in an improved fashion. Mr Robert Hesketh of Meels, also instituted some drainage, but as he and the other landowner worked independently, nothing important was accomplished, and the Mere was still flooded in wet seasons. In 1849. Sir T Dalrymple Hesketh, by means of a steam engine lifted the water from the Rufford portion of the Mere, and permanently laid dry 800 acres. Mr Charles Scarisbrick Esq., of Scarisbrick Hall, who in 1843 had acquired the Bold moiety of North Meels for the sum of £132,000, and in 1848 a portion of the Hesketh moiety, cut drains and also erected a steam engine (1853) at Crossens to lift water from the lowest level of his land. This answered well for a long time, but as year by year, more area was added to the low level, it became necessary to reconstruct the Crossens pumping station, and three of the finest centrifugal pumps in the world were put down in 1882. The larger pump lifts 70 tons or 18,000 gallons per minute, and the two smaller ones each 45 tons per minute making a total lift of 42,000 gallons per minute and thus the task was effectively accomplished. Lord Lilford and Lord Derby did not join in the work of the Crossens pumping station, and their land yet floods in the winter and is then used for skating recreation. It may here be notice that the peat gradually decays under cultivation, and the level of the ground gradually lowers, and may make fresh demands for increased pumping power and increased outfall. United action, would however, greatly reduce the expense, and would easily cope with all difficulties. In old days the Douglas, when overcharged, then flooded the Mere, but the action of the Croston Drainage Board does away with this danger, and what is wanted now is that the Southport Corporation and the landowners should unite and make a good outfall for the sluice and connect it with the Bon Hole channel. The late Charles Scarisbrooke brought all of the water from his estates to Crossens to help scour the channel and evil has resulted from this supply being diminished. He also desired to make a cutting connecting the canal with a port at Crossens.

As might be expected, a waterway on the Mere, was a great disideratum to early settlers, and so on its shores we find Holmes Wood Hall, Ruf-

*This 17th century map of Lancashire,
shows Martin Mere quite clearly*

ford Hall, Tarlscough Hall, Hurleston Hall, Burscough Priory, Martin Hall, Scarisbrick Hall and I would even include Gorsuch Hall, Rannacres Hall and Leigh House. It is true that these houses were a little distance from the ordinary water's edge but then the shore varied with the season and the rainfall, and it was necessary that the buildings be above the floodline. Those residing on the Mere suffered from ailments owing to the damp situation, but they enjoyed fowling and fishing, and also social intercourse. They moved about a great deal, as the numerous signatures to charters, and deeds testify. The Mere being connected with the river Douglas, skiffs could journey to Parbold, Tarleton, Eccleston, Croston,

Hoole and even Longton, and we find the names of the lords of these manors appended to the Scarisbrick charters.

Leigh House on Mere Brow stands on the site of an older dwelling which belonged to the Bannisters of Bank Hall. The present building is an interesting one owing to its thick walls, and to a cattle fair, sanctioned by a charter in 1700 having been held there.

The Bannisters were mighty magnates until their power was broken by Thomas, Earl of Lancaster, in 1315. Then at Preston, Adam Bannister won one fight, but his troops being called upon to encounter a second army at Maudlands, they were defeated, and Adam and Sir William Lea were beheaded on Leyland Moor, being betrayed by one with whom they sought concealment. In 1555 Wm. Bannaster [sic] was found possessed of 60 messages, 80 acres of arable land, 20 acres of meadow, 100 acres of pasture, and 500 acres of moss in North Holmes (Mere Brow), in the vill. of North Meols, for which he paid a "ferm" of 2s., to the Duchy. At that time Mere Brow was in the Parish of Croston, and previously the Abbey of Crockersand had had lands there.

Holmeswood Hall, on the way to Rufford, is situated on an eminence overlooking the Mere. The present building has walls four ft thick on a stone foundation. The stone foundation tells of a previous building built of timber and plaster. In a gable wall of the barn are to be seen the wheatsheaf of the Fittons of Rufford and the double eagle of the Heskeths of Hesketh. There is also a date which tells of another building which was erected by Thomas Hesketh, who was knighted in 1553 and was a rave warrior, as well as a wealthy and liberal man. In the old maps, a fenced park is shown, extending far to the north of the house, the road till recently running on its south face. But even before the maps were issued Holmes was famous for its Arthurian fights, and to this Arthur's pit and Wigan Lane testify. The hall chapel stood on the north side of the building, and near to Arthur's pit. The Mere fishery was much esteemed by the owner of the hall, and we find the retainers of Sir Thomas Hesketh capturing the nets of Henry Banastre, of Legh House, in 1557, and assaulting him and his attendants because it was thought that they had trespassed. This led to another free fight and a lawsuit at Westminster.

Rufford Old Hall stands in an angle formed by the junction of Martin Mere with the River Douglas. Its distinguishing feature is the grand banqueting Hall, probably built by Sir Thomas Hesketh, who died in 1523.

288

This picturesque, half-timbered structure is surrounded by an elegant louver lantern, which allowed the smoke to escape in the early days when the fire was built up in the centre of the hall. At the top end is a dais with canopy, and above it a secret hiding chamber in the gable. At the lower end of the hall where the kitchen is approached, is a magnificent screen or "spear", marvellous for its design and solidity. The east wing, built in brick by Sir Thomas Hesketh in 1662, contains many antique and beautiful articles of furniture. As in the church there is a marble slab to the memory of Sir Thomas Hesketh, who died in 1363, we may think there was an edifice before the present structure, although I think that the first edifice would be the manor house of the Fittons (where they never resided), and their home at Holmeswood Hall, and so were accounted lords also of Hesketh and Rufford societies. Rufford New Hall was erected in 1798, and again a Thomas Hesketh was the builder. The Preston and Liverpool road was deserted, and many cottages were pulled down in order to form the park surrounding the hall.

Burscough Priory is about seven miles from Southport, and some of its ruined arches can be seen from the L and Y Railway between Burscough and Ormskirk. stations.. It was founded in 1190 and was endowed with the Manor of Martin, with a right to the fisheries of Martin Mere. Ormskirk Church belonged to the Priory, and was the parish church of the dwellers on the south side of the Mere, even as Croston was the parish church for the residents on its north shore. The names of many of the lake dwellers in the year 1366 are to be found amongst the 476 subscribers to the Chantry of Our Lady in Ormskirk Church. A straight road, now partially obliterated, ran from the Priory to the church, and on the demolition of the Priory, its stones and bells were taken along this road for the erection of the great tower at Ormskirk. In 1539 there were five religious men at the priory and 42 servants. Its land afterwards came into the possession of the Earls of Derby, whose ancestors, The Lathoms, originally founded the Priory, and much of the Mere land yet belongs to the Derby family.

The Priors of Burscough were well capable of looking after their Mere privileges, and we find that when in the reign of Edward II, Prior John leased a holme or island, known as Blake-ness (Black Nose), in his water at Merton Mere for 19 years to Thurstan de Northleigh, he made a stipulation that only those should go upon the island who had permission from the Prior. Blakenase was only 90ft long by 20ft wide, and the rent was one penny, paid at Easter. The island could only have been useful for sporting purposes, and Thurston would need a canoe to visit it.

Martin Hall is on the rising ground above New Lane Station.. The place is called Battle Holme and the name goes back to Arthurian days. The present building is a fine old mansion, and it passed in 1890, from the Dicconson family of Wrightington to the Earl of Derby. and stand on the site of what I think was a moated grange belonging to Burscough Priory. A straight, good road connects the two buildings. This is called Lord's-Lane and the name implies that it was on demesne land. It has been noted that there were 32 servants at the Priory, and they would work the adjacent land. At the dissolution of the Priory, the hall passed into the hands of the Wrightingtons, who leased it to the Breres of Preston. The Elizabethan Hall (now used as a kitchen) has a stone mullioned window about 16 feet long. Its heraldry belongs to John Brere and Katharine Walton, and is dated 1614. About 1682, James Starkey of Marton re-sided here. It then becomes a farm house and was occupied by the Holcrofts. Thomas Holcroft (1745-1809), the great dramatist, author of "The Road to Ruin", came of this family. The place is haunted but I do not know what form the boggart takes, though I have conversed with those who have seen it. When I went up the fine old staircase I noticed an empty bedroom, and judged that there was the "Ghost's Walk". Strange lights have shone from the attic window across the Mere, but I think the room (only reached by a ladder), must have been a priest's hole in times of trouble.

Hurleston Hall, lying on the north side and about a quarter of a mile from the Ormskirk road is a fine old half-timbered mansion and is well worthy of a visit. The manor is mentioned as *Hirletun* and *Hiretun* in the Domesday Book and included in the adjoining manors of Scarisbrick, Merton and Burscough. The site of the hall has been occupied from Saxon times and it is an elevated knoll bordered by a brook and was originally noted and to some extent fortified.

There is a well in an underground room of the present building. A survival from very ancient days and which would have supplied the garrison with water. The building is now a farmhouse, but the medieval Great Hall and its timbered minstrel's gallery (the latter now partitioned into bedrooms), can easily be made out. I enquired about its reputed boggart but was told that rent day is now the only boggart time. The manor house must once have been much larger and I judge a chapel must have been included among the buildings for about 1870 the remains were moved of an old burial ground and were re-interred in Ormskirk churchyard.

Hurleston Hall now forms part of the Scarisbrick Hall estate but in the old days had its resident landowner. His name constantly appears as witness to the Scarisbrick charters. In 1270 Richard who leases lands in Depedale to Quenilda, daughter of his brother Simon of Schawe, and speaks of his brother Henry Lord of Scarisbrick and yet calls himself 'son of' Edusa Hurton. In 1427 Elizabeth, widow of Gilbert of Hurton releases Henry Gilbert, William and Thomas Scarisbrick, from law proceedings for the death of her husband. In 1529 there was a dispute between Thomas Scaresbrek and Humphrey Hurleton about a piece of land called Little Grandearth, bounded by Hurleton Brook and they appropriated arbitrators. Humphrey, naming the Prior of Burscough and Henry ffarington; Thomas naming Thomas and Bartholemew Hesketh and a £100 bond was given that they would all abide by the decision as issued by Richard Halsall the legal advisor.

Scarisbrick Hall has ever been the chief residence on the banks of Martin Mere. The first recorded resident was a dependent or scion of the Latham family and in like fashion the residence at Hurleston Hall and Gorsuch Hall were dependent on the Scarisbrick family and received all their education at Scarisbrick Hall. The original hall I conceive to have been erected where there is now a moat in the grounds and I think it was a timber and plaster erection. About the year 1200 the Lord of Scarisbrick assisted in endowing the new Prior at Burscough. Again, in 1260 another gift was made to the Priory. The connection with the Latham family was shown by Henry Scarisbrick witnessing 1369 the will of Thomas of Latham, leaving a legacy of 2 marks for the repair of three bridges at Parbold. Then a delph was discovered and worked at Scarsbrick a manor house of stone was begun on the present site and the stream was dammed and a lake formed so that a water mill might obtain power and the family might have a supply of fish at hand for their fast.

In 1420 and 1447 licenses were obtained from the Bishop of Lichfield that service might be performed in the Hall Chapel and in the will of Thomas Scarisbrick (1530), mention is made of two vestments and chasubles, two albs, a chalice, a corporal, altar, super alter, twelve images in box cases, and two images not closed up. Two mass books are also mentioned and these, with the duplicate vestments, show that mass was performed with some dignity, and that two chaplains were often present. Besides the chapel, the house contained a large hall furnished with tables and trestles and benches which was the general living room, and where the dependants had their meals in day time and slept at night. There would also be some chambers where the Scarisbricks and the Bower-Maidens slept. And also an attic for the dairy and other maids. The offices mentioned in the will are kitchen, butlery, larder house, and brewhouse, oxhouse, and furniture to us would seem to us very scant and poor.

Edward Scarisbrick (1566-1599) rebuilt or rather restored the house, and over the eastern porch he placed the inscription "Edwarde Scarisbricke anno domini 1595". His initials E.S and the date 1569 were also placed on a marble chimney piece. I wondered at finding in a catholic house pictures of Henry VIII and his wives, but they would likely be placed on the wainscote by Thomas Scarisbrick while Edward would afterwards add those of Queen Mary, Queen Elizabeth, Mary Queen of Scots, and Darnley.

As might be expected, the Scarisbricks, like the Heskeths and the Blundells, adhered to and supported the older form of religion. The priests have left us a continuing series of books forming what is called St Mary's Library, and the signatures in them form an unbroken list (chiefly belonging to the society of Jesus), of clergy who ministered at the hall during two centuries and a half. These works are of a devotional character and the only novel included among them is *L'Historie d'Heliodore* dated Paris 1559 and contain the signature "Henry Ecclestone December 10th 1584". Father Henry Scarisbrick YS.I , after officiating for nine years departed hastily on the flight of James II in 1688 leaving his pocket book behind him. It showed that he received 2s 6d or 5s for saying mass with intention and 4s for weddings. For expenditure there is wige 5s; shooses 3s 6d; penknife 6d; picktooth 1s 6d; cravat 2s; gune £1 1s 6d, coat £3 6s. Father Scarisbrick disappeared but I don't think he went far from his birth place, and finally in Lancashire at the close of 1701.(sic)

To give a history of the family would entail a long and confusing list of names. Scarisbrick Manor in the Domesday Book was included in Hurlestone but Henry Scarisbrick, who lived about 1360 got a verdict that the Hurlestone owner was his vassal and that he had the right of guardianship to the son and the disposing of him in marriage which was worth about £20 to the superior lord. About Cromwell's time, Hugh Worthington, a tenant of the estate, found a number of Roman coins. This Worthington resided at Heaton's Bridge in 1665. Robert Scarisbrick got into trouble in the rebellion of 1715, but delivering himself up he was let out on bail. He consoled himself with sporting, racing and perhaps a little smuggling. In the 18th Century the Gorsuch family succeeded the Ecclestone property and there came to an end. Basil Thomas Scarisbrick succeeded in the property and took the name of Ecclestone and died there aged 73 in 1789 although he had succeeded to Scarisbrick Hall.

Charles Scarisbrick, whose name is so well known in Southport, ever since he brought a moiety of the North Meals estate in 1843 for £132,000 was originally called Ecclestone but retook the old name of Scarisbrick when he obtained possession of the estate in 1835. At one time he was the richest commoner in England and his income was over £100,000 per annum. He was thus enabled to carry out his schemed for the draining of Martin Mere and to rebuild Scarisbrook Hall. He enlisted the aid of A.W.Pugin, but both architect and patron died before the completion of the present building and then the work was carried out by his sister, Lady Hunloke who assumed the name Scarisbrick and was aided by E.W.Pugin.

The existing hall at Scarisbrick is somewhat cramped by being built into the site of the previous hall. Some of the outbuildings of which still remain. It is a noble and costly structure especially characterised by a wealth of detail and exquisite finish. It reminds one in its style of the Houses of Parliament, and of the King Edward VI Grammar School at Birmingham. It is a lovely structure enshrined amongst woods and gardens and having a beautiful lake and heronry in front. Its carving, gilding and pinnacles make it a perfect dream in stone, but I confess that its vast cost might have erected a more commodious and yet equally noble house. I think the Great Hall (45 feet by 30 feet) is the chief feature of the mansion. It is galleried and its height to the top of the lantern is eighty feet. Another striking feature of the interior is the beauty and amount of old oak carving. This was imported from the Netherlands and most visitors pause before the Christ crowned with thorns, the Resurrection morn and the Jesse tree. To be brief I will just mention the outsized turreted gateway, the bronze figures from the exhibition of 1851, and which their purchaser would not part with to Prince Albert, and the exquisite dairy erected at the old mill dam at a cost of £2000. Lady Scarisbrick's daughter married the Marquis of Casteja who was of English ancestry and from him the estate was descended to the present Count in whose noble line it is likely to remain. The estate extends over about 14,000 acres and its rent roll has been returned at £27,000.

293

Scarisbrick village is adorned with some comfortable almshouses and with a Catholic chapel. The chapel is worth inspection especially for the handsome carved pulpit. In its grounds there is a noticeable cross which marks the resting place of the late Charles Scarisbrick Esq. He died in May 1860 and the coffin was carried in a straight line from the Hall to the resting place. He left property worth £3 million and deserves to take high rank as a scientist, and as an intelligent, benevolent landowner.

Gorsuch Hall, is midway between Scarisbrick Hall and Halsall. Adam de Gosefordistche lived here about 1190, was son of Walter de Scaresbreeke and his descendants occupied the ancestral hall until the eighteenth century when they removed to Ecclestone. The hall was burned down in 1816 but gives its name to an adjacent farm. Near this spot is the base of a farmhouse boundary cross which gave the name to a family which intermarried with the Gorsuch and Scarisbrick families and assisted in the reclaiming of Martin Mere. In the latter part of the 13th Century Walter de Gosfordsyke obtained twelve acres of meadow called the Wyke, from Alan de Meles,. These were claimed by John Bold Lord of North Meols, who in 1554 forcibly carried away the crops. The Gorsuch family must ultimately have obtained possession for in law records (1642) Edward Gorsuch was found in possession of 12 acres of meadow, four acres of pasture and 20 acres of moss, situated in North Meols. The possession of Wyke carried with it the right to certain fisheries which were much appreciated and caused several law suits. In 1353, some Rufford men took 20s worth of bream from "Lee Wyk" and it was decided that the fisheries of the Mere belonged to the Lords of Rufford and North Meols, to the Abbot of Cokersand, and to the Prior of Burscough. "Pro in diviso".

The records of the Gorsuch family, judging from the fragments that we have, would be interesting. Thus there is (dates 1291), an agreement for a marriage between two infants; Robert, son of Walter of Gosfordesiche and Agnes, Grandchild of William Bird of Donnington, in which William binds himself to pay a dowry of ten marks, or 133 shillings and 4 pence, in three instalments. Child marriage was frequent and money was scarce. This was not a happy marriage, for thirteen years after Robert was slain in Church Street, Ormskirk. William of the Cross, a dyer, smote him with a staff or "dodgestpade" on the left side of the head and was tried for murder in 1305. Homicide was not uncommon; thus in 1332 Walter of Gosfordesiche and Adam his son, were charged with the murder of Adam, Abbot of Burscough Priory, and were tried and acquitted in 1333. A later tragic incident was when John, son of Edward Gorsuch in 1679, at the bidding of his enamorata, betrayed Fr Penketh S.J.

Space is too brief, or one might tell of the family at Renacres dating back to 1200: The family at Wyke dating from 1180: of the Cross Family, named from the cross near Gorsuch Hall: of the Meremen who fought at Crecy in 1346 and of those who accompanied Sir Edward Scarisbrick to Agincourt in 1415. Meremen to, were involved in the fight of the Banister family at Preston in 1315, and in the downfall of the Holland family (1400) of Upholland. The merrymakings relieved the troubles as the espousal of William of Hurelton and Maud of Gosfordesiche in 1402 and the marriage in 1405 of Robert Halsale and Ellen of Scaresbrec when a dowry of 200 marks was paid. Funerals, too were pageants, when a Scarisbrick was buried at Burscough and Ormskirk and stopped at the wayside cross.

Another sad thing was the existence of slavery. Walter of Scaresbrecke in 1260 gave a slave and his chattels to Godiva, his daughter.

On the old maps of Martin Mere, three islands are marked. I have recovered the names of two viz. Whassam Hyle, and Otterhouseholme. Quassame or Whassum was marked by a mill called Whawshaw Mill. The mill is in the centre of Martin Mere on Lord Derby's land and can be seen from the L & Y Railway. It is now disused though at one time so popular that men now living had to wait all night for their turn to get batches ground. Being on an island, when evil spirits were cast out, they were sent to Wholesone (present pronunciation) Brow. In 1240, Roger of Hurleton let the island to John of Crossens (Crosens) for a rent of 24 pence payable half yearly at Martinmass and Pentecost. In 1300 Walter, son of John of Quassam, gave the island to Gilbert son of Henry, lord of Scarisbrec. In 1336 Richard del Cross, gave the said Gilbert , who then had a windmill, certain approaches to it. Otterhouseholme is mentioned in 1440 when Gilbert of Scarisbrek makes it over to his two chaplains Edward Banastre and William Becansawe. Clay Brow may have been the third island of the maps. A small island in a deed of 1492 is called Fishery of Wyke and another Doe Hyle. There may be noticed also in the Deeds detached pieces of Martin Mere called Black Lache and Merton Pool. Other deeds tell of disputes about estate boundaries, for inhabitants being few these were not well defined on the wastelands. Great credit is due to the late Charles Scarisbrick and the Scarisbrick Trustees for their spirited and continuous efforts for the drainage of Martin Mere so that whereas the Mere brought no rent to the Lords, it now brings in a rent proportionate to that held in 1290 where arable land was valued at 12 shillings per acre and paid a yearly rent of one shilling per acre.

Tarlscough was given by the Lathoms to Burscough Priory. Its old title was Terlescough Wood and it was in this wood that a boy was discovered by Lord Lathom in an Eagle's nest. The boy was adopted by Lady Lathom and the incident was commemorated by the crest of the eagle and child.

ꞏ GLOSSARY ꞏ

mere – Old English word: lake.

Ordnance Survey – The UK's official survey organisation that produces large-scale maps of the country.

dimpsey – archaic English word meaning gloaming, twilight.

About the author...........

Jonathan Downes was born in Portsmouth in 1959, and spent much of his childhood in Hong Kong where, surrounded by age-old Chinese superstitions and a dazzlingly diverse range of exotic wildlife, he soon became infected with the twin passions for exotic zoology and the paranormal which were to define his adult life. He spent some years as a nurse for the mentally handicapped but began writing professionally in the late 1980s. He has now written over twenty books. He is also a musician and songwriter who has made a number of critically acclaimed but commercially unsuccessful albums.

In 1992 he founded The Centre for Fortean Zoology, with the aim of co-ordinating research into mystery animals, bizarre and aberrant animal behaviour and his own particular love of zooform phenomena (paranormal entities which only appear to be animals!)

He has searched for Lake Monsters at Loch Ness, pursued sea serpents and the grotesque Cornish owlman—which inspired his most famous book *The Owlman and Others* - chased big cats across westcountry moorland, and in 1998 and 2004 went to Latin America in search of the grotesque vampiric Chupacabra. He is a popular public speaker both in the UK and the United States, where he regularly appears at conventions talking about his many expeditions and his latest research projects.

He lists his hobbies as Tequila, radical politics, the music of Scott Walker, books and more books. He is Leo with Scorpio Rising and believes that Harpo Marx is the funniest man to have ever lived. He is divorced and commutes between his house in Exeter where he lives with two cats, a dog with asymmetrical ears, and a monitor lizard called Roger, and his old family home in rural North Devon.

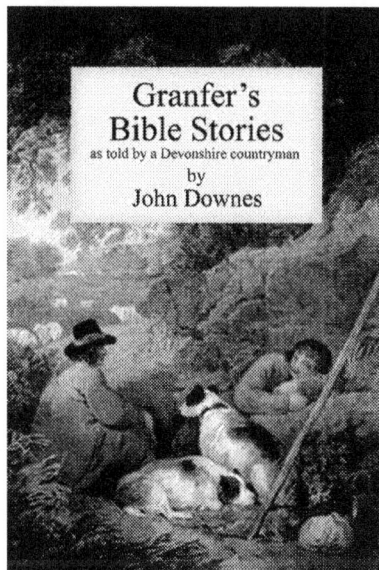
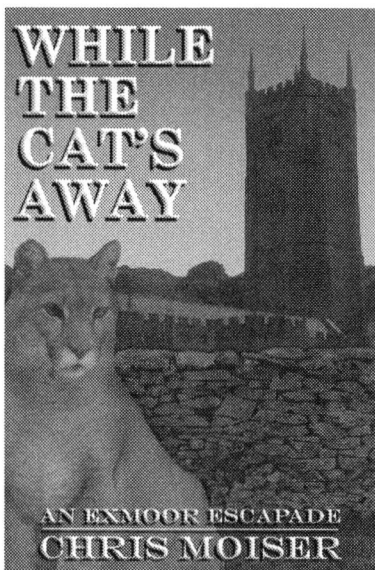

www.ingramcontent.com/pod-product-compliance
Lightning Source LLC
Chambersburg PA
CBHW060329200326
41519CB00011BA/1886